高等学校计算机专业教材精选·算法与程序设计

C程序设计
实验实践教程

刘玉英 主编
肖启莉 邹运兰 编著

清华大学出版社
北京

内 容 简 介

学习 C 语言的目的在于掌握编写程序的技能,实验与实践是学习 C 语言的必备环节,基于这一点,本书内容包括三部分:实验指导、课程设计以及《C 语言程序设计——案例驱动教程》一书的习题参考答案。

实验指导部分包含 18 个实验,实验题目类型包括填空题、改错题、程序阅读题、编程题。每个实验中都以一个具体题目为例,讲解如何分析问题、编写代码、实验步骤、运行程序、分析程序运行结果是否正确。课程设计部分包含 8 个项目,有游戏设计,如"俄罗斯方块"、"五子棋"等,也有小系统设计,如"英汉电子词典"等。每个项目都给出算法分析、难点提示以及部分代码,要求读者根据给出的提示信息编写部分代码。

本书可以与《C 语言程序设计——案例驱动教程》配套使用,作为 C 语言课程的实验教材和课程设计教材,也可以单独使用。

本书封面贴有清华大学出版社防伪标签,无标签者不得销售。
版权所有,侵权必究。侵权举报电话: 010-62782989 13701121933

图书在版编目(CIP)数据

C 程序设计实验实践教程/刘玉英主编. —北京:清华大学出版社,2013.2 (2015.12 重印)
高等学校计算机专业教材精选·算法与程序设计
ISBN 978-7-302-30543-9

Ⅰ. ①C… Ⅱ. ①刘… Ⅲ. ①C 语言-程序设计-高等学校-教材 Ⅳ. ①TP312

中国版本图书馆 CIP 数据核字(2012)第 257598 号

责任编辑:	张 民 谢 琛 薛 阳
封面设计:	傅瑞学
责任校对:	白 蕾
责任印制:	王静怡

出版发行: 清华大学出版社
  网　　址: http://www.tup.com.cn, http://www.wqbook.com
  地　　址: 北京清华大学学研大厦 A 座　　　　邮　编: 100084
  社 总 机: 010-62770175　　　　　　　　　　　邮　购: 010-62786544
  投稿与读者服务: 010-62776969, c-service@tup.tsinghua.edu.cn
  质量反馈: 010-62772015, zhiliang@tup.tsinghua.edu.cn
  课件下载: http://www.tup.com.cn, 010-62795954
印 装 者: 虎彩印艺股份有限公司
经　　销: 全国新华书店
开　　本: 185mm×260mm　　　印 张: 18.5　　　字　数: 449 千字
版　　次: 2013 年 2 月第 1 版　　　　　　　　　印　次: 2015 年 12 月第 2 次印刷
印　　数: 3001~3200
定　　价: 29.50 元

产品编号: 047093-01

出 版 说 明

我国高等学校计算机教育近年来迅猛发展,应用所学计算机知识解决实际问题,已经成为当代大学生的必备能力。

时代的进步与社会的发展对高等学校计算机教育的质量提出了更高、更新的要求。现在,很多高等学校都在积极探索符合自身特点的教学模式,涌现出一大批非常优秀的精品课程。

为了适应社会的需求,满足计算机教育的发展需要,清华大学出版社在进行了大量调查研究的基础上,组织编写了《高等学校计算机专业教材精选》。本套教材从全国各高校的优秀计算机教材中精挑细选了一批很有代表性且特色鲜明的计算机精品教材,把作者们对各自所授计算机课程的独特理解和先进经验推荐给全国师生。

本系列教材特点如下。

(1) 编写目的明确。本套教材主要面向广大高校的计算机专业学生和理工科学生,使学生通过本套教材,学习计算机科学与技术方面的基本理论和基本知识,接受应用计算机解决实际问题的基本训练。

(2) 注重编写理念。本套教材作者群为各校相应课程的主讲,有一定经验积累,且编写思路清晰,有独特的教学思路和指导思想,其教学经验具有推广价值。本套教材中不乏各类精品课配套教材,并力图努力把不同学校的教学特点反映到每本教材中。

(3) 理论知识与实践相结合。本套教材贯彻从实践中来到实践中去的原则,书中的许多必须掌握的理论都将结合实例来讲,同时注重培养学生分析、解决问题的能力,满足社会用人要求。

(4) 易教易用,合理适当。本套教材编写时注意结合教学实际的课时数,把握教材的篇幅。同时,对一些知识点按教育部教学指导委员会的最新精神进行合理取舍与难易控制。

(5) 注重教材的立体化配套。大多数教材都将配套教师用课件、习题及其解答,学生上机实验指导、教学网站等辅助教学资源,方便教学。

随着本套教材陆续出版,相信能够得到广大读者的认可和支持,为我国计算机教材建设及计算机教学水平的提高,为计算机教育事业的发展做出应有的贡献。

清华大学出版社

前　言

　　C语言程序设计是一门实践性很强的课程，除了需要掌握必要的理论知识外，还要经过足够的实践训练，才能真正掌握编写程序的技能并为后续课程的学习做好知识储备，否则，学习的效果将大打折扣。实践训练包括两方面的内容：课程实验和课程设计。课程实验可以使理论知识得以验证和应用，从中学习编写程序的基础技能和调试程序的技巧，掌握基础算法知识。课程设计可以将所学知识进行综合性应用，进一步加深对所学知识的理解和应用。

　　本书基于在课程实验和课程设计方面给学习者以指导和引导，包括三方面的内容：实验指导、课程设计以及《C语言程序设计——案例驱动教程》一书的习题参考答案。

　　在实验指导部分共有18个实验，囊括了C语言中的全部知识点，题型包括程序填空、程序改错、程序阅读及编程题，题目范围广、趣味性强、题量大，任由读者自由选择。从内容上可划分为三个模块，基本数据类型及其运算、数据的输入与输出、三种控制结构程序设计为第一模块，数组、字符串、函数以及编译预处理为第二模块，指针、结构、文件为第三模块，每个模块都有一个综合性实验，其中的实验题目难度有所提升，覆盖前面的知识点。每个实验都以一个具体的题目为例，讲解如何分析问题、编写代码，演示上机实验步骤直至运行程序、分析结果是否正确，引导读者步步深入。每个实验后面设有思考题，希望读者完成一个实验后能对实验中遇到的知识点或问题有一个归纳总结。

　　在课程设计部分设计了8个训练项目，有游戏设计的，如"五子棋"、"贪吃蛇"、"俄罗斯方块"等，也有小系统设计的，如"通讯录"、"英汉电子词典"、"运动会成绩统计与管理"、"火车订票系统"、"图书信息管理系统"等。每个项目都给出算法分析与流程图、难点提示，给出部分源代码，读者可以在此基础上编写未给出的部分模块的代码。由于游戏设计中用到C标准库函数中的画图函数，所以该部分内容是基于TC编译系统的，其余部分基于Visual C++ 6.0编译系统。

　　在习题参考答案中，给出了《C语言程序设计——案例驱动教程》一书中全部习题的参考答案。尤其要说明的是，对于编程题，求解一个问题，编写的程序代码不是唯一的，只要解题思路正确，程序运行结果满足题意要求即可。

　　本书由刘玉英给出写作大纲和基础要求，实验指导部分由刘玉英、肖启莉和邹运兰共同完成，课程设计部分由肖启莉和邹运兰共同完成，《C语言程序设计——案例驱动教程》一书习题参考答案部分由刘玉英、肖启莉和刘臻共同完成，最后由刘玉英统编定稿。

　　限于作者水平，书中难免存在错误和不妥之处，恳请各位读者、教师、专家批评与指正。

作　者
2012年10月

目 录

第一部分 实 验

实验一 编写并运行简单的 C 程序 …… 3
实验二 基本数据类型及其运算 …… 10
实验三 数据的输入与输出 …… 16
实验四 选择结构程序设计 …… 26
实验五 循环结构程序设计 …… 37
实验六 综合实验（一） …… 48
实验七 一维数组与二维数组 …… 54
实验八 字符串 …… 63
实验九 函数（一） …… 72
实验十 函数（二） …… 84
实验十一 变量的存储类型与生存期 …… 96
实验十二 编译预处理与位运算 …… 102
实验十三 综合实验（二） …… 108
实验十四 指针与字符串 …… 117
实验十五 结构与联合 …… 128
实验十六 排序与查找程序设计 …… 140
实验十七 文件操作 …… 153
实验十八 综合实验（三） …… 164

第二部分 课 程 设 计

项目一 通讯录 …… 175
项目二 五子棋游戏 …… 189
项目三 英汉电子词典 …… 199
项目四 运动会成绩统计与管理 …… 204
项目五 俄罗斯方块 …… 211
项目六 火车订票系统 …… 222
项目七 图书信息管理系统 …… 234
项目八 贪吃蛇游戏 …… 245

第三部分 习题参考答案

第1章 C 程序知识初步 …… 253
第2章 基本数据类型及其操作 …… 254

第3章　选择结构程序设计……………………………………………………………………257
第4章　循环结构………………………………………………………………………………261
第5章　数组……………………………………………………………………………………265
第6章　函数……………………………………………………………………………………269
第7章　指针……………………………………………………………………………………273
第8章　结构及其他……………………………………………………………………………278
第9章　文件……………………………………………………………………………………281
第10章　编译预处理与位运算…………………………………………………………………285
参考文献…………………………………………………………………………………………287

第一部分

实验

第一辑

血書

实验一 编写并运行简单的 C 程序

一、实验目的与要求

1. 熟悉 C 语言程序的集成开发环境。
2. 熟悉 C 语言上机过程,即从源程序编辑、编译、连接、运行并查看结果的过程。
3. 通过运行简单的 C 程序,初步了解 C 程序的基本结构与书写规则。
4. 了解 printf 函数的基本功能。

二、实验内容

1. 输入下列程序并对其进行编译,观察屏幕上显示的编译信息。如果出现"出错信息",找出原因并改正,再进行编译;如果无错,则进行连接。若编译、连接无误,运行程序,观察运行结果。

(1)

```
#include<stdio.h>
void main()
{   printf("Hello!How are you!\n");}           /*输出一行字符信息*/
```

(2)

```
#include<stdio.h>
void mian()
{   printf("**********\n");
    printf("Welcome!\n");
    printf("**********\n");
}
```

2. 分析下列程序的运行结果,并上机调试运行,验证自己的分析结果。

(1)

```
#include<stdio.h>
void main()
{   printf("*\n");
    printf("**\n");
    printf("***\n");
    printf("****\n");
    printf("*****\n");
}
```

(2)

```
#include<stdio.h>
void main()
{   int a,b,sum;                              /*定义3个整型变量*/
    a=25;                                     /*为整型变量赋值*/
    b=50;
    sum=a+b;                                  /*求两个整型变量之和,并赋给sum*/
    printf("The sum of %d and %d is %d.\n",a,b,sum);
}
```

3. 编写程序

(1) 编程显示输出如图1.1所示的图形。
(2) 编程显示输出如图1.2所示的字母图形。
(3) 编写程序,显示输出如图1.3所示的信息。
(4) 在屏幕上显示输出学生的学号、姓名和班级等信息,结果如图1.4所示。

```
    *              *     *       1. 中文
   **              *     *       2. 英语
  ***              ******        3. 日语         学号          姓名          班级
 ****              *     *       4. 法语         ——————————————————————
*****              *     *       5. 俄语         201001486    张兵          计算机02班

  图 1.1           图 1.2          图 1.3                     图 1.4
```

(5) 已知某位学生的语文、数学和英语的成绩分别为85分、90分、94分,求该学生3门课的平均成绩。

三、实验步骤

作为本教程的第一个实验,首先要了解在 Visual C++ 编程环境下,如何实现 C 程序的编辑、编译、连接、运行程序以及查看运行结果。下面以一个具体的题目为例,介绍运行一个 C 程序的基本步骤。

1. 题目:分析下列程序,写出程序的运行结果,并上机验证。

```
#include<stdio.h>
void main()
{   printf("Hello world!\n"); }                /*输出一行信息*/
```

2. 上机调试

对于该问题,通过分析可知程序的功能是输出一行字符信息(即 Hello world!)。下面通过上机进行验证,具体步骤如下:

(1) 启动 VC++。执行"开始"→"所有程序"→Microsoft Visual Studio 6.0→Microsoft Visual C++ 6.0 命令,进入 VC++ 编程环境,如图1.5所示。

(2) 新建文件。执行 File→New 命令,单击 Files 选项卡,如图1.6所示,先在 File 文本

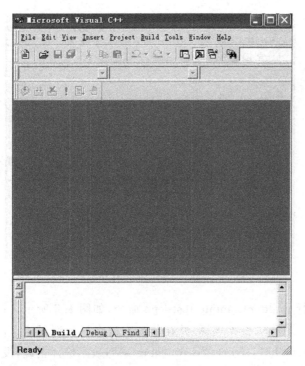

图 1.5　VC++ 窗口

框中输入 test，把 C 语言源程序文件命名为"test.cpp"，在 Location 下拉列表框中选择已经建立的文件夹，如"C：\program；"然后选中 C++ Soure Files 选项，单击 OK 按钮，即在"C:\program"下新建了文件"test.cpp"，并显示编辑窗口和信息窗口，如图 1.7 所示。

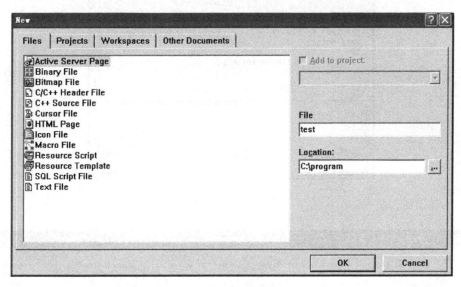

图 1.6　新建文件

（3）编辑和保存。在编辑窗口中输入源程序，如图 1.7 所示。然后执行 File→Save 命令，保存源文件。

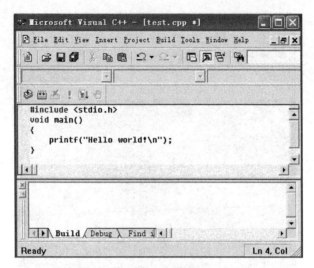

图1.7 编辑源程序

(4)编译。执行Build→Compile test.cpp命令,如图1.8所示。在弹出的消息框中单击"是"按钮,如图1.9所示,开始编译,并在信息窗口中显示编译信息,如图1.10所示。信息窗口中出现的"test.obj-0 error(s),0 warning(s)",表示编译正确,没有发现错误和警告,并生成了目标文件test.obj。

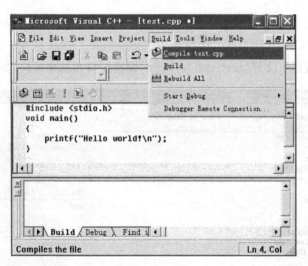

图1.8 编译源程序

图1.9 产生工作区消息框

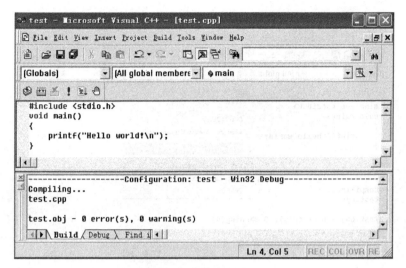

图 1.10　程序编译后信息显示窗口

如果显示错误信息,说明程序中存在严重错误,必须改正;如果显示警告信息,说明这些错误并未影响目标文件的生成,但通常也应按照系统给定的提示信息进行改正。

(5) 连接。执行 Build→Build test.exe 命令,开始连接,并在信息窗口中显示连接信息,如图 1.11 所示。

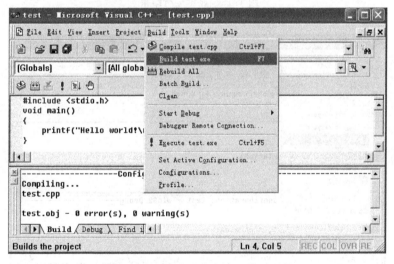

图 1.11　连接成功并产生可执行文件

信息窗口中出现的"test.exe-0 error(s),0 warning(s)"表示连接成功,并生成了可执行文件 test.exe。

(6) 运行。执行 Build→! Execute test.exe 命令,如图 1.12 所示,自动弹出运行窗口,如图 1.13 所示,显示运行结果"Hello world!"。其中,Press any key to continue 是系统提示用户按任意键退出运行窗口,返回到 VC++ 编辑窗口。

(7) 关闭程序工作区。执行 New→Close Workspace 命令,如图 1.14 所示。在弹出的对话框中单击"是"按钮,关闭工作区,如图 1.15 所示。

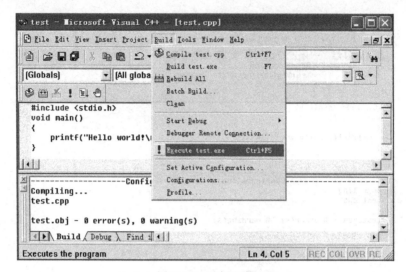

图 1.12 运行程序

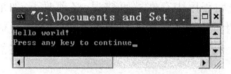

图 1.13 显示运行结果

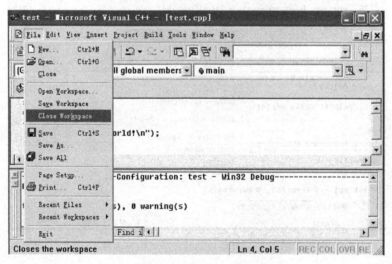

图 1.14 关闭工作区

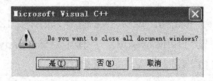

图 1.15 关闭所有文档窗口

当一个程序执行完毕,需要进行一个新程序的编辑、编译、连接、执行等过程时必须关闭当前工作区,建立新工作区,即从步骤(2)开始,重复该过程。

(8)打开文件。如果要再次打开 C 源文件,可以执行 File→Open 命令,在文件夹"C：\program"中选择文件 test.cpp；或者在文件夹 C：\program 中直接双击文件 test.cpp。

(9)查看 C 源文件、目标文件和可执行文件的存放位置。经过编辑、编译、连接和运行后,在文件夹"C：\program"(如图 1.16 所示)和"C：\program\Debug"(如图 1.17 所示)中存放着相关文件。其中,源文件"test.cpp"在文件夹"C：\program"中,目标文件"test.obj"和可执行文件"test.exe"都在文件夹"C：\program\Debug"中。

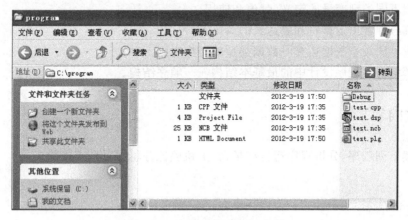

图 1.16　文件夹"C：\program"

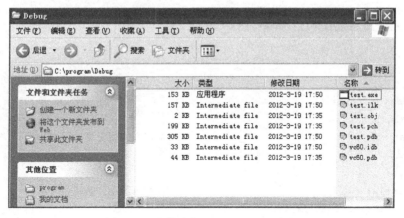

图 1.17　文件夹"C：\program\Debug"

四、思考题

1. 上机调试一个 C 程序具体分为哪几个步骤？
2. 一个 C 程序的结构是由哪些部分构成的？
3. 编写 C 语言程序时,每条语句的结束符是什么？

实验二　基本数据类型及其运算

一、实验目的与要求

1. 了解 C 语言中数据类型分类，掌握 C 语言的基本数据类型。
2. 掌握不同类型变量的定义、赋值和使用。
3. 掌握各种常用运算符和表达式的应用。
4. 掌握不同类型数据运算时数据类型的转换规则。
5. 了解结构化程序设计中的最基本结构——顺序结构。

二、实验内容

1. 阅读下列程序，分析程序运行结果，并上机验证分析结果是否正确。
(1)

```
#include<stdio.h>
void main()
{   int x=1,y=2;                           /*定义整型变量 x、y*/
    printf("x=%d y=%d sum=%d\n",x,y,x+y);
    printf("10 Squared is: %d\n",10*10);
}
```

(2)

```
#include<stdio.h>
void main()
{   float x;                               /*定义实型变量 x*/
    int i;
    x=3.6;
    i=(int)x;                              /*数据类型强制转换*/
    printf("x=%f,i=%d',x,i);
}
```

(3)

```
#include<stdio.h>
void main()
{   char ch;                               /*定义字符型变量 ch*/
    int i;
    float f;
    double d;
    ch='A';
```

```
    i=1;
    f=2.0;
    d=2.5;
    printf("%f", ch/i+f*d-(f+i));
}
```

2. 填空题

请根据题意在下面各程序中划线处填写适当的语句或表达式,使之能够运行并获得正确的结果。

(1) 下面程序的输出结果是 16.00。

```
#include<stdio.h>
void main()
{   int a=9, b=2;
    float x=_____,y=1.1,z;
    z=a/2+b*x/y+1/2;
    printf("%5.2f\n",z);
}
```

(2) 下面程序的功能是:输出整数 2580 的后两位数值 80。

```
#include<stdio.h>
void main()
{   int  x=2580,y;
    y=_____;
    printf("%d\n",y);
}
```

(3) 已知一个圆的半径为 1.5cm,求该圆的周长和面积。

```
#include<stdio.h>
#define PI 3.14                    /*定义符号常量 PI*/
void main()
{   float r,l,s;
    _____;
    l=2*PI*r;
    s=PI*r*r;
    printf("l=%f,s=%f\n",l,s);
}
```

(4) 现有算术表达式 ab/4.6c+8%d,求该表达式的值并输出。其中,a=5,b=9,c=7,d=21。

```
#include<stdio.h>
void main()
{   _____;
    a=5,b=9,c=7,d=21;
    t=_____;
```

```
        printf("t=%f\n",t);
}
```

3. 改错题

下列各程序中存在两处或两处以上的错误,请仔细阅读程序,根据题意进行修改,使程序能够正确运行。

(1) 该程序功能:已知两个实数 a 和 b,求它们的和并输出。

```
#include<stdio.h>
void main()
{   int   a=4,b=8;
    t=a+b;
    printf("%f\n",t);
}
```

(2) 该程序功能:已知两个整数 a 和 b,求它们的平均值并输出。

```
#include<stdio.h>
void main()
{   int   a,b=18;
    float t;
    t=(a+b)/2;
    printf("%f\n",t);
}
```

(3) 该程序功能:已知变量 a 和 c 的值,进行简单的算术运算后输出其结果。

```
#include<stdio.h>
void main()
{   int a,b;
    float c;
    a=5;   c=5.7;
    b=(a*3.8+c)%6;
    printf("%d\n",b);
}
```

(4) 求华氏温度 100°对应的摄氏温度。计算公式为:c=5/9(f−32),其中:c 表示摄氏温度,f 表示华氏温度。

```
void main()
{   int c;f;
    f=100;
    c=5/9(f-32);
    printf("f=d,c=%d\n",f,c);
}
```

(5) 已知物品的单价,根据数量 x 的值求其总金额。

```
#include<stdio.h>
#define PRICE 30
```

```
void main()
{   float x=5;
    PRICE=PRICE * x;
    printf("%f   %f\n",x,PRICE);
}
```

4. 编程题

(1) 编写程序,已知:a=7,x=2.5,y=4.7(a 为整型,x、y 为浮点型),求算术表达式 x+a%3*(int)(x+y)%2/4 的值。

(2) 编写程序,要求用变量初始化的方法使 c1、c2 这两个变量的值分别为 97、98,然后分别按整型和字符型输出。

(3) 编写程序,实现任意两个整数的交换。例如,刚开始 a=3、b=4,交换以后 b=3、a=4。

(4) 当 n 为 593 时,分别求出 n 的个位数字(digit1)、十位数字(digit2)和百位数字(digit3)的值(提示:n 的个位数字 digit1 的值是 n%10,十位数字 digit2 的值是(n/10)%10,百位数字 digit3 的值是 n/100)。

三、实验步骤

本次实验要求学生学会正确地实现有关数据的描述和简单的算术操作。

1. 题目:编写程序,已知:a=2,b=3,x=3.9,y=2.3(a、b 为整型,x、y 为浮点型),求算术表达式(float)(a+b)/2+(int)x%(int)y 的值。

2. 算法分析

任何程序都包含两方面内容:数据的描述和数据的操作。其中数据的描述是为数据操作服务的,数据操作是为了实现问题的求解。为了有效计算题中表达式的值,根据题意首先定义 a、b 为 int 型,x、y 为 float 型。此外,还要再定义一个用来保存表达式结果的变量 z。很明显,z 的数据类型应该为浮点型 float。接下来就是有关数据操作部分:首先给变量 a、b、x、y 赋值,分别为 2、3、3.9、2.3;然后计算表达式(float)(a+b)/2+(int)x%(int)y,并将该表达式的值赋值给变量 z。最后将变量 z 的值输出即可。

在表达式中 x、y 进行求余运算,根据运算符"%"的运算规则,运算对象必须为整型数,而 x、y 都是浮点数,所以必须将 x、y 强制转换为整型。

3. 根据分析,写出代码如下:

```
#include<stdio.h>
void main()
{   int a,b;
    float x,y,z;
    a=2;    b=3;
    x=3.9;   y=2.3;
    z=(float)(a+b)/2+(int)x%(int)y;
    printf("(float)(a+b)/2+(int)x%(int)y =%f\n",z);
}
```

4. 上机调试

(1) 编辑源程序：

将源程序代码输入到 Visual C++ 代码编辑窗口中，如图 2.1 所示。

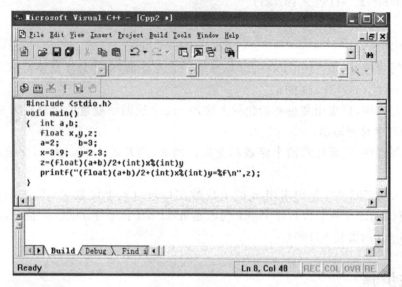

图 2.1　编辑源程序

(2) 编译、连接：

单击菜单 Build→Build，进行编译和连接，如图 2.2 所示。编译、连接中若发现错误将在输出窗口中显示，如图 2.3 所示。

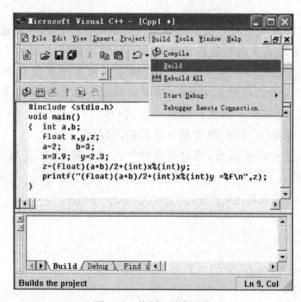

图 2.2　编译、连接程序

观察输出窗口，发现编译、连接无误，严重错误(errors)0，警告错误(warnings)0，可执行文件已经生成。

图 2.3 编译、连接后信息显示窗口

（3）运行程序：

单击菜单 Build→! Execute Cpp1.exe,或者使用键盘组合键 Ctrl+F5,也可以直接单击工具栏中的按钮 ! 运行程序,输出结果如图 2.4 所示。从图中可以看出,算术表达式 (float)(a+b)/2+(int)x%(int)y 的值为 3.500000。

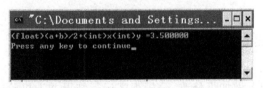

图 2.4 运行程序窗口

四、思考题

1. C 语言中基本数据类型有哪几种？
2. 算术运算符求余"％"使用时有什么特殊要求？
3. 运算符的优先级和结合性对表达式的运算结果有影响吗？
4. 在什么情况下需要使用强制类型转换？

实验三 数据的输入与输出

一、实验目的与要求

1. 掌握使用 scanf 函数实现数据输入的方法。
2. 掌握使用 printf 函数实现数据输出的方法。
3. 掌握 getchar 函数和 putchar 函数的使用。
4. 熟悉顺序结构程序设计的方法。
5. 掌握转义字符的使用。

二、实验内容

1. 分析下列程序,并上机验证程序运行结果。

(1)

```
#include<stdio.h>
void main()
{   char c1,c2;
    c1='A';
    c2='a';
    printf("c1=%c, c2=%c\n",c1,c2);
}
```

(2) 在题(1)的基础上做如下修改,分析程序运行结果并上机验证。

① 在语句 printf("c1=%c, c2=%c\n",c1,c2);后增加:printf("c1=%d, c2=%d\n",c1,c2);

② 把语句 char c1,c2;改为:int c1,c2;

③ 语句 c1='A'; c2='a';改为:c1=65; c2=97;

(3)

```
#include<stdio.h>
void main()
{   short i;                        /*定义短整型变量 i*/
    i=-4;
    printf("i:dec=%d,oct=%o,hex=%x,unsigned=%u\n",i,i,i,i);
}
```

(4)

```
#include<stdio.h>
void main()
{   float f=3.1415927;
```

```
    printf("%f,%5.4f,%3.3f",f,f,f);
}
```

(5)

```
#include<stdio.h>
void main()
{   char c='x';
    printf("c:dec=%d,oct=%o,hex=%x,ASCII=%c\n",c,c,c,c);
}
```

(6)

```
#include<stdio.h>
void main()
{   char c1='b',c2,c3;
    c2=getchar();
    putchar('c');    putchar('\n');
    putchar(c1);     putchar('\n');
    putchar(c2);     putchar('\n');
    c3=c1-32;
    c1=c3+32;
    putchar(c1);     putchar('\n');
    putchar(c2);     putchar('\n');
    putchar(c3);     putchar('\n');
    putchar('\101'); putchar('\n');
}
```

(7)

```
#include<stdio.h>
void main()
{   printf("\t*\n");
    printf("\t\b***\n");
    printf("\t\b\b*****\n");
}
```

(8)

```
#include<stdio.h>
void main()
{   printf("\102   \x43   D\n");
    printf("E\b=\n");
    printf("I say:\"How do you do?\'\n");
    printf("\\C Program\\\n");
    printf("Turbo \'C\'");
}
```

2. 填空题

请根据题意在下面各程序中划线处填写适当的语句或表达式,使之能够运行并获得正确的结果。

(1) 从键盘输入两个整数,计算它们的和并输出。

```
#include<stdio.h>
void main()
{   int m,n,t;
    scanf(_____);
    t=a+b;
    printf(_____);
}
```

(2) 以下程序的功能是实现美元兑换人民币计算,假设美元与人民币的汇率是 1 美元兑换 6.27 元人民币。

```
#include<stdio.h>
void main()
{   double rmb,dollar;
    printf("Enter RMB:");
    scanf("_____",&rmb);
    dollar=_____;
    printf("RMB%.2lf can exchange dollar %0.2lf\n",_____);
}
```

(3) 从键盘输入长方形的长和宽,求长方形的面积和周长并输出,保留 2 位小数。

```
#include<stdio.h>
void main()
{   double length,width,perimeter,area;
    printf("Enter length of rectangle:\n");
    _____;
    printf("Enter width of rectangle:\n");
    _____;
    perimeter=2*(length+width);
    area=length*width;
    _____;
}
```

(4) 从键盘输入 3 个整数 a、b、c,求它们的平均值并按如下形式输出:

average of **,** and **is **.**。
```
#include<stdio.h>
void main()
{   int a,b,c;
    _____;
    scanf("_____",&a,&b,&c);
    t=(a+b+c)/3.0;
    printf(_____);
}
```

(5) 使用 getchar() 函数输入一个字符,用 printf() 输出;用 scanf() 函数输入一个字符,用 putchar()函数输出。

```
#include<stdio.h>
void main()
{   char c;
    printf("input the first char:\n");
    _____;
    printf("_____",c);
    printf("input the second char:\n");
    scanf("%c",_____);
    _____;
}
```

3. 改错题

下列各程序中存在两处或两处以上的错误,请仔细阅读程序,根据题意进行修改,使程序能够正确运行。

(1) 下面程序的功能为：从键盘上输入变量 a、b 的值,计算 c＝a＊b 并输出。

```
#include<stdio.h>
void main()
{   int a,b,c;
    scanf("%d,%d",a,b);
    c=a*b
    printf("%d");
}
```

(2) 下面程序的功能为：输入圆的半径 r,计算圆的周长和面积。

```
#include<stdio.h>
void main()
{   float r,c,s,p;
    p=3.141593;
    scanf("%f", r);
    c=2pr;
    s=pr²;
    printf("c=%f s=%f\n",c,s);
}
```

(3) 下面程序的功能为：计算某个数 x 的平方,并分别以"y＝x＊x"和"x＊x＝y"的形式输出 x 和 y 的值。例如,假设 x 的值为 3,则输出信息如下：

9=3＊3
3＊3=9

```
#include<stdio.h>
void main()
{   int y;
    y=x*x;
    printf("%d=%d*%d",x);
    printf("d*%d=%d",y);
```

}
(4)

```
#include<stdio.h>
void main()
{   int m,n;
    float a,b;
    scanf("%d%f",&m,&a);
    scanf("%d%f",&b,&n);
    printf("%f\n",m/n+a);
    printf("%f\n",m%b+n);
}
```

4. 编程题

（1）编写程序，从键盘任意输入一个字符，求其前导字符和后续字符并输出。例如，字符 b 的前导字符是 a，后续字符是 c。

（2）已知 a=4,b=2.2,c=51274,c1='A'，编写程序得到如下输出结果：

a=3,b=1.2,c=-14262
c1='A' or 65

（3）编写程序，使用 getchar 和 putchar 两个函数，从键盘接收三个字符 YOU，输出这三个字符的小写 you。

（4）编写程序，输入一个正整数，分别按十进制、八进制和十六进制输出该数。

（5）从键盘输入两个整数 a 和 b，计算并按照以下格式输出它们的和、差、积、商和余数。例如，假设 a=7,b=3，则输出结果如下：

7+3=10
7-3=4
7*3=21
7/3=2
7%3=1

（6）从键盘输入实数 x，计算并按照以下格式输出函数 y=2x+6 的值。例如，假设输入实数 x 为 2.5，则输出结果如下：

y=f(2.5)=11.000000

三、实验步骤

本实验主要目的是使学生学会正确使用 scanf 和 printf 函数实现数据的输入和输出。

1. 题目：下面程序的功能为：从键盘输入变量 a、b 的值，计算 c=a*b 并输出，程序中存在错误，改正错误并上机运行。

```
#include<stdio.h>
void main()
```

```
{   int a,b;
    scanf("%d,%d",a,b);
    c=a*b
    printf("%d",c);
}
```

2. 上机调试

(1) 编辑源程序：

将源程序代码输入到 Visual C++ 代码编辑窗口中，如图 3.1 所示。

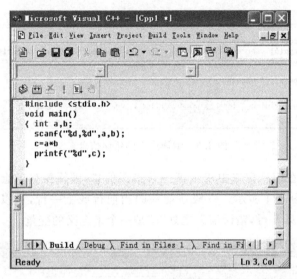

图 3.1　编辑源程序

(2) 编译：

单击菜单 Build→Compile 进行编译，如图 3.2 所示。信息窗口中显示编译错误信息，如图 3.3 所示。编译过程发现严重错误(errors)2，警告错误(warnings)0。

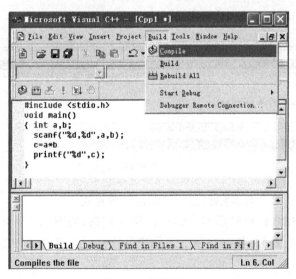

图 3.2　编译程序

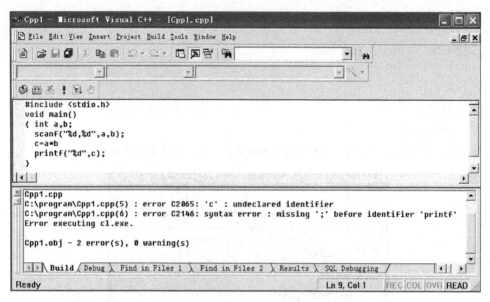

图 3.3 编译产生的错误信息

(3) 找出错误。在信息窗口中双击第一条错误信息,编辑窗口就会出现一个箭头指向程序出错的位置,如图 3.4 所示。一般在箭头的当前行或上一行,可以找到出错语句。如图 3.4 所示箭头指向第 5 行,错误信息之处"c"是一个未定义的变量。

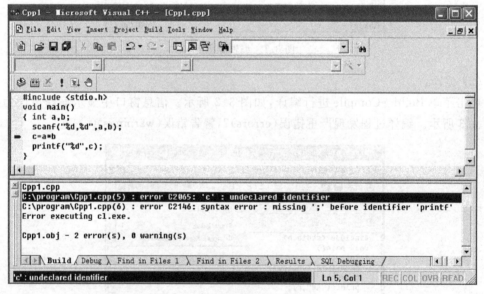

图 3.4 编译错误对应的位置

(4) 改正错误。在第 3 行,将对变量"c"的定义加进去。

(5) 重新编译。信息窗口显示本次编译的错误信息,如图 3.5 所示。双击该错误信息箭头指向出错位置,错误信息指出在该语句前缺少分号。改正错误,在该语句的上一条语句末尾补上一个分号。

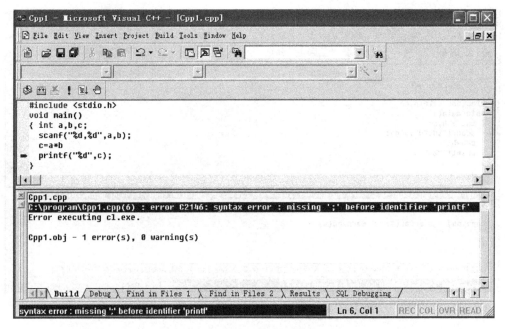

图 3.5 重新编译后产生错误信息以及位置

(6) 再次编译。发现 0 error(s),2 warning(s),也就是说还有 2 个警告错误,如图 3.6 所示。双击第一条错误信息,箭头指向程序的第 4 行,如图 3.7 所示,错误信息指出变量 a 没有初始化,同时发现第二条信息是说变量 b 也没有初始化。该错误造成的原因在于输入语句 scanf("％d,％d",a,b);中变量 a、b 前缺少符号 &,意味着从键盘输入的数据不能保存在系统为它们分配的内存中,在接下来的运算中要使用它们的值,也就无法找到。

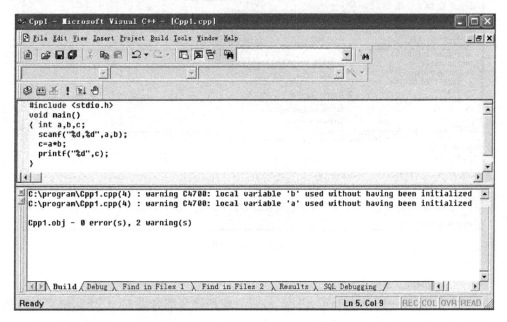

图 3.6 第 3 次编译后产生的错误信息

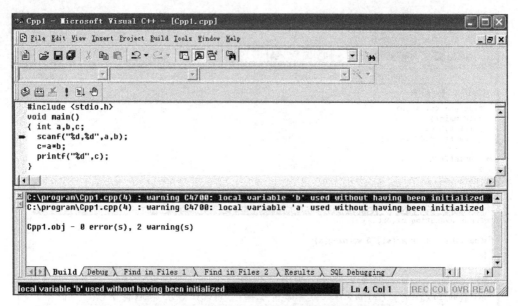

图 3.7　第 3 次编译产生的一个错误所在的位置

(7) 再次改正错误并重新编译。信息窗口显示编译正确。

(8) 连接。执行菜单 Build→Build Cpp1.exe 进行连接,如图 3.8 所示。连接结果如图 3.9 所示,在信息窗口中,发现 0 error(s),0 warning(s),也就是说连接成功。在连接过程中,若发现错误将在信息窗口中显示。

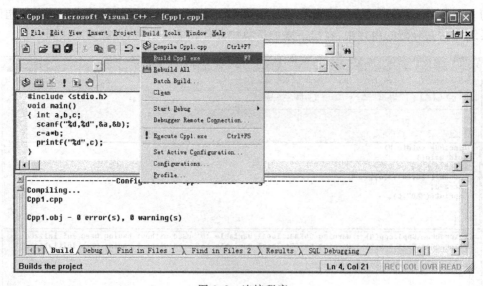

图 3.8　连接程序

(9) 运行程序：

单击菜单 Build→! Execute Cpp1.exe,或者使用键盘组合键 Ctrl＋F5,也可以直接单击工具栏中的按钮！运行程序,输出结果如图 3.10 所示。从图中可以看出,当输入 3,5 后按回车键,输出的结果是 15,与题目要求的结果一致,按任意键返回。

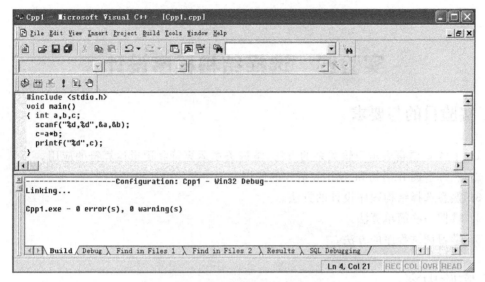

图 3.9　连接后产生的信息

图 3.10　程序运行结果输出窗口

四、思考题

1. %d、%3d、%-5d、%.5f、%8.2f、%-8.2f、%5.3s、%10.2e 分别表示什么含义。
2. 使用 scanf 函数输入数据时应注意什么。
3. 输入字符时,需要输入单引号吗？输入字符串时要输入双引号吗？连续输入多个字符时需要间隔符吗？若要输出'A'或"abc",输出控制格式应如何设置？
4. 设有输入语句 scanf("%d%c",&x,&ch); 对应以下三组输入时,输出结果一样吗？

① 2↙

② 2a↙

③ 2　a↙

实验四 选择结构程序设计

一、实验目的与要求

1. 了解 C 语言表示逻辑关系的方法,掌握关系运算符和逻辑运算符的应用。
2. 熟练掌握 if 语句和 switch 语句的使用。
3. 熟悉选择结构程序设计的方法。
4. 掌握一些简单算法。
5. 学习调试程序的方法。

二、实验内容

1. 分析下列程序运行结果,并上机验证。

(1)

```c
#include<stdio.h>
void main()
{   int m=5;
    if(m++>5) printf("%d\n",m);
    else printf("%d\n",m--);
}
```

(2)

```c
#include<stdio.h>
void main()
{   int a,b,c,s=0,w=0,t=0;
    a=-1; b=3; c=3;
    if(c>0) s=a+b;
      if(a<=0)
      {   if(b>0)
            if(c<=0) w=a-b;
      }
      else if(c>0) w=a-b;
    else t=c;
    printf("%d %d %d\n",s,w,t);
}
```

(3) 下列程序连续执行 4 次,分析当分别输入 1、2、3、4 时的结果。

```c
#include<stdio.h>
void main()
```

```
{  int  n;
   scanf("%d",&n);
   switch (n)
   {  case 1:
      case 2:putchar('n');
      case 3:printf("%d\n",n); break;
      default: printf("Good!\n");
   }
}
```

(4)

```
#include<stdio.h>
void main()
{  int x=1,a=0,b=0;
   switch(x)
   {  case 0:b++;
      case 1:a++;
      case 2:a++;b++;
   }
   printf("a=%d,b=%d\n",a,b);
}
```

(5)

```
#include<stdio.h>
void main()
{  int a=3,b=6,c=9;
   a=a>b?a:b;
   a=a>c?a:c;
   printf("%d \n",a);
}
```

2. 填空题

请根据题意在下面各程序中划线处填写适当的语句或表达式,使之能够运行并获得正确的结果。

(1) 从键盘输入两个数,输出其中的较小者。

```
#include<stdio.h>
void main()
{  int a,b,min;
   scanf("%d%d",_____);
   if(_____)min=a;
   else _____;
   printf("min=%d\n",min);
}
```

(2) 以下程序的功能是从键盘输入3个整数,并按由大到小的顺序输出。

```
#include<stdio.h>
void main()
{  int a,b,c,t;
   scanf("%d%d%d",&a,&b,&c);
   if(a<b) { t=a; a=b; b=t; }
   _____
   _____
   printf("%d,%d,%d",a,b,c);
}
```

(3) 从键盘输入一个字符,如果该字符是字母,输出"char!";如果该字符是数字,输出"digit!";否则输出"other!"。

```
#include<stdio.h>
void main()
{  char ch;
   scanf("%c",&ch);
   if(_____) printf("char!\n");
   else if(_____) printf("digit!\n");
   _____ printf("other!\n");
}
```

(4) 任意输入三条边 a、b、c 后,若能构成三角形且为等腰、等边和直角,则分别输出 DY、DB 和 ZJ,若不能构成三角形则输出 NO。

```
#include<stdio.h>
void main()
{  float a,b,c,a2,b2,c2;
   scanf("%f%f%f",&a,&b,&c);
   printf("%5.1f,%5.1f,%5.1f",a,b,c);
   if(a+b>c&&b+c>a&&a+c>b)
   {  if(_____)printf("DY");
      if(_____)printf("DB");
      a2=a*a;b2=b*b;c2=c*c;
      if(_____)printf("ZJ");
      printf("\n");
   }
   else printf("NO\n");
}
```

(5) 输入圆的半径 r 和运算标志字符 m,按照运算标志符进行计算。当输入的标志字符为 a 时,计算圆的面积;当输入的标志字符为 b 时,计算圆的周长;当输入的标志字符为 c 时,则二者都计算。

```
#include<stdio.h>
#define PI  3.14              /*定义符号常量PI*/
```

```
void main()
{   float r,s,c;
    char m;
    printf("请输入圆的半径 r 和运算标志字符 m: \n");
    scanf(_____);
    if(_____)
    {   s=PI*r*r;
        printf("area is %7.2f",s);
    }
    if(_____)
    {   c=2*PI*r;
        printf("circle is %7.2f",c);
    }
    if(_____)
    {   s=PI*r*r;
        c=2*PI*r;
        printf("area and circle are %7.2f,%7.2f",s,c);
    }
}
```

(6) 现有如下函数，根据自变量 x 的值计算出相应 y 的值。

$$y = \begin{cases} 0 & x < 0 \\ x & 0 \leqslant x < 10 \\ 10 & 10 \leqslant x < 20 \\ -0.5x + 20 & 20 \leqslant x < 40 \\ -2 & x \geqslant 40 \end{cases}$$

```
#include<stdio.h>
void main()
{   int x,c;
    float y;
    scanf("%d",&x);
    if(_____)c=-1;
    else c=_____;
    switch(c)
    {   case -1:y=0; break;
        case 0:y=x; break;
        case 1:y=10; break;
        case 2:
        case 3:y=-0.5*x+20; break;
        default:y=-2;
    }
    if(_____) printf("y=%f",y);
    else printf("error\n");
}
```

3. 改错题

下列各程序中存在两处或两处以上的错误,请仔细阅读程序,根据题意进行修改,使程序能够正确运行。

(1) 下列程序是输入实数 x,计算并输出下列分段函数 f(x)的值,保留 1 位小数。

$$f(x) = \begin{cases} 1/x & x = 10 \\ x & 其他 \end{cases}$$

```
#include<stdio.h>
void main()
{   double x;
    printf("Enter x:");
    scanf("%f",x);
    if(x=10)   y=1/x;
    else (x!=10) y=x;
    printf("f(%.2f)=%.1f\n",x,y);
}
```

(2) 以下程序的功能是:从键盘输入一个整数,求其绝对值并输出。

```
#include<stdio.h>
void main()
{   int x;
    scanf("%d",x);
    if(x>=0) y=x;
    else  (x<0)   y=-x;
    printf("%d",y);
}
```

(3) 以下程序的功能是:输入三个整数,输出三个整数值为中间的数。

```
#include<stdio.h>
void main()
{   int x,y,z,;
    printf("enter the numbers:\n");
    scanf("%d%d%d", &x,&y,&z);
    if(x<y<z) printf("middle is %d\n",y);
    else if(y<x<z) printf("middle is %d\n",x);
    else  printf("middle is %d\n",z);
}
```

(4) 以下程序的功能是:从键盘输入两个整数,按照由大到小的顺序输出。

```
#include<stdio.h>
void main()
{   int a,b;
    scanf("%d%d",&a,&b);
```

```
    if(a<b)
    t=a;
    a=b;
    b=t;
    printf("%d%d\n",a,b);
    else printf("%d%d\n",a,b);
}
```

(5) 以下程序的功能是：判断输入的一个整数是否能同时被 5 和 7 整除，若能整除，则输出 yes，否则输出 no。

```
#include<stdio.h>
void main()
{   int t;
    scanf("%d",t);
    if(x%5==0||x%7==0) printf("yes\n");
    else printf("no\n");
}
```

(6) 以下程序的功能是：从键盘输入 3 个整数，输出其中的最小数。

```
#include<stdio.h>
void main()
{   int a,b,c,t,min;
    scanf("%d%d%d",&a,&b,&c);
    t=(a<b)?b:a;                    /*条件运算符的应用*/
    min=(t>c)?t:c;
    printf("min=%d\n",min);
}
```

(7) 以下程序的功能是：从键盘输入一个成绩，判断该成绩是否通过考试。

```
#include<stdio.h>
void main()
{   int grade;
    scanf("%d",&grade);
    switch(grade/10)
    {   case 10,9,8,7,6: printf("pass!");
        default: printf("not pass!");
    }
}
```

4. 编程题

(1) 输入整数 a 和 b，若 a＋b 大于 100，则输出 a＋b 百位以上的数字，否则输出两数之和。

(2) 输入一个英文字母，判断该字母是大写字母还是小写字母。

（3）用 switch 实现：
$$y = \begin{cases} -1 & (x < 0) \\ 0 & (x = 0) \\ 1 & (x > 0) \end{cases}$$

（4）输入两个字符，若这两个字符的 ASCII 码之差为偶数，则输出它们的后继字符，否则输出它们的前驱字符。

（5）从键盘输入两个整数，当第一个数为 1 时，计算并输出第二个数除以 3 的余；否则，计算并输出第二个数除以 5 的余。

（6）输入某个点 A 的平面坐标 (x,y)，判断点 A 是在圆内、圆外还是在圆周上，其中圆心坐标为 $(2,2)$，圆的半径为 1。

（7）编程计算与日历有关的问题：
① 输入年、月，输出该月的天数。
② 给出年、月、日，计算出该日是该年的第几天。
③ 2011 年元旦是星期六，问 2011 年 10 月 1 日是星期几。

（8）输入一个整数，将与该整数相对应的月份的英语名称输出。例如，输入 1，输出 January。请分别使用 if 语句和 switch 语句实现。

（9）输入一个整数，判断它能否被 3、5、7 整除，并根据情况输出下列信息：
① 能同时被 3、5、7 整除。
② 能同时被 3、5、7 中的两个数整除，并输出这两个数。
③ 只能被 3、5、7 中的一个数整除，并输出这个数。
④ 不能被 3、5、7 中的任何一个数整除。

（10）企业发放的奖金根据利润提成。当利润低于或等于 10 万元时，奖金提 10%；当利润大于 10 万元小于等于 20 万元时，低于 10 万元的部分按 10% 提成，高于 10 万元的部分按 7.5% 提成；当利润大于 20 万元小于等于 40 万元时，高于 20 万元的部分可提成 5%；当利润大于 40 万元小于等于 60 万元时，高于 40 万元的部分可提成 3%；当利润大于 60 万元小于等于 100 万元时，高于 60 万元的部分提成 1.5%，高于 100 万元时，超过 100 万元的部分按 1% 提成。从键盘输入当月利润，分别用 if 语句和 switch 语句计算应发放奖金。

三、实验步骤

本实验要求掌握选择结构的实现方法，其中常见的实现方法是使用 if 语句实现单分支和二分支，利用 switch 语句实现多分支。

1. 题目：输入 x，计算并输出下列分段函数 $f(x)$ 的值，改正下列程序中的错误。

$$y = f(x) = \begin{cases} 0 & (x = 0) \\ 1/x & (x \neq 0) \end{cases}$$

```
#include<stdio.h>
void main()
{   double x,y;
```

```
        printf("Enter x:");
        scanf("%lf",x);
        if(x!=0) y=1/x
        else y=0;
        printf("f(%.2f)=%.2f\n",x,y);
}
```

2. 上机调试

(1) 编辑源程序,单击 ❀(Compile)按钮,出现的第一条编译错误信息是:

`missing ';' before 'else'`

双击该错误信息,箭头指向 else 所在的行,错误信息指出在 else 的上一行缺少分号。补上分号后,重新编译,新出现的第一条错误信息是:

`"local variable 'x' used without having been initialized"`

双击该错误信息,箭头指向 scanf 所在的行,错误原因是 x 的前面少了 &。补上 & 后,重新编译并连接,都正确。

(2) 执行 Tools→Customize 命令,自动弹出 Customize 对话框,如图 4.1 所示。单击对话框中的 Toolbars 选项,如图 4.2 所示,选中 Debug 复选框,则在 VC++ 窗口中显示调试工具条,如图 4.3 所示。

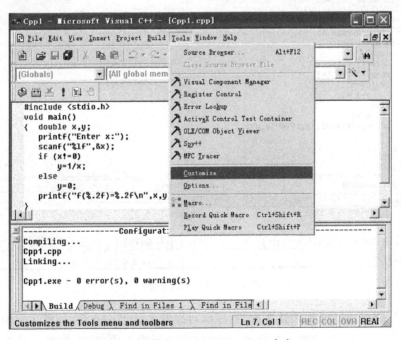

图 4.1　执行 Tools→Customize 命令

(3) 程序调试开始,单击调试工具条中的 ❀(Step Over)按钮,每次执行一行语句,如图 4.4 所示,编辑窗口中的箭头指向某一行,表示程序将要执行该行。图 4.4 中列出了变量窗口(Variables Window)和观察窗口(Watch Window),在观察窗口中可以改变变量的值。

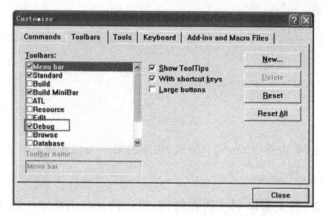

图 4.2 选择工具栏

图 4.3 调试工具条

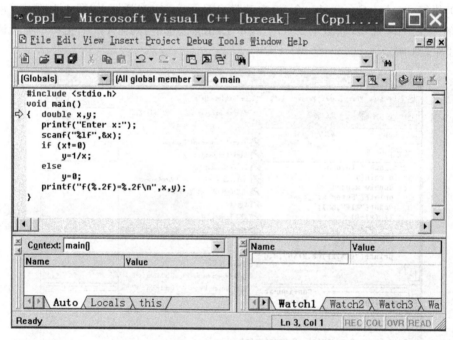

图 4.4 程序调试开始

(4) 单击 (Step Over)按钮 3 次,程序执行到输入语句这一行,如图 4.5 所示,同时运行窗口显示"Enter x:",如图 4.6 所示,此时将要执行输入语句,继续单击 按钮,在运行窗口中输入 10,如图 4.7 所示,按下回车键后,箭头指向"if(x!=0)"这一行,如图 4.8 所示。在变量窗口中可以看到变量 x 的值为 10.000000000000。

(5) 继续单击 (Step Over)按钮 2 次,箭头指向"else"这一行,如图 4.9 所示。在变量窗口中可以看到变量 y 的值为 0.10000000000000。

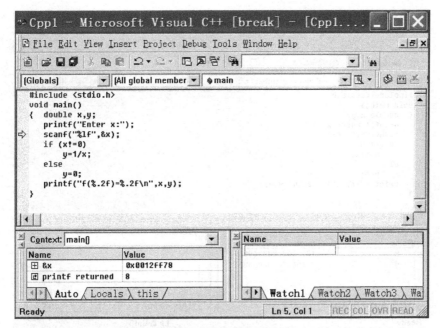

图 4.5 程序单步调试

图 4.6 运行窗口

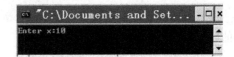

图 4.7 在运行窗口中输入变量的值为 10

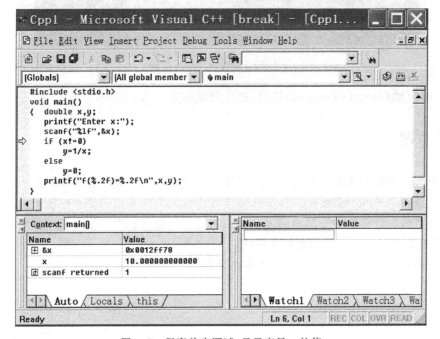

图 4.8 程序单步调试,显示变量 x 的值

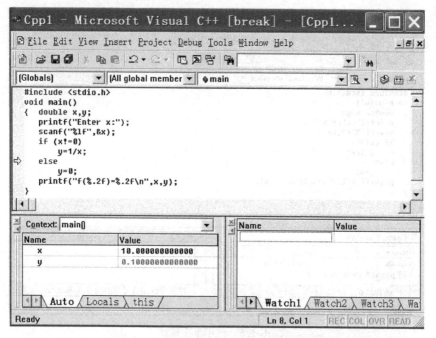

图 4.9　程序单步调试,显示变量 y 的值

(6) 继续单击 (Step Over)按钮 2 次,运行窗口显示运行结果如图 4.10 所示,符合题目的要求。

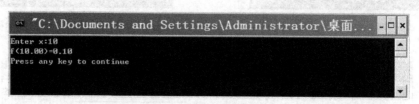

图 4.10　运行窗口显示结果

(7) 单击 (Stop Debugging)按钮,程序调试结束。

四、思考题

1. 逻辑运算符和关系运算符的应用有什么区别?其优先级的高低如何?
2. 运算符＝＝和＝的区别是什么?在进行条件判断时常用的是＝＝还是＝?
3. 浮点数在计算机中不能精确表示。在测试浮点数的相等时不要使用＝＝运算符,而要测试它是否在某个浮点值的差值范围内。
4. 举例说明什么是单分支、二分支和多分支。

实验五 循环结构程序设计

一、实验目的与要求

1. 熟练掌握 while 语句、do-while 语句和 for 语句的用法。
2. 掌握 break 语句和 continue 语句的使用,以及它们之间的区别。
3. 熟悉循环结构程序设计的方法。
4. 掌握常用算法,如穷举法、迭代法和递归法等。
5. 进一步掌握调试程序的方法。

二、实验内容

1. 阅读下列程序,分析程序的运行结果,并上机验证分析结果是否正确。

(1)

```
#include<stdio.h>
void main()
{   int n=4;
    while(n--) printf("%d ",--n);
}
```

(2)

```
#include<stdio.h>
void main()
{   int a,s,n,count;
    a=2;s=0;n=1;count=1;
    while(count<=7)
    {   n=n*a; s=s+n; ++count;   }
    printf("s=%d",s);
}
```

(3)

```
#include<stdio.h>
void main()
{   int x=6,y=7;
    printf("%d,",x++);
    printf("%d\n",++y);
}
```

(4)
```
#include<stdio.h>
void main()
{   int a=10,x=8,y=11;
    x=y++;
    y=--x;
    a=x+++y;
    printf("%d,%d,%d\n",a,x,y);
}
```

(5)
```
#include<stdio.h>
void main()
{   int a=10,x=8,y=11;
    x=y++;
    y=--x;
    a=x+++y;
    printf("%d,%d,%d\n",a,x,y);
}
```

(6)
```
#include<stdio.h>
void main()
{   int a,i,s;
    i=1;a=0;s=1;
    do
    {   a=a+s*i; s=-s; i++;
    }while(i<=10);
    printf("a=%d",a);
}
```

(7) 当输入为字符串"qwert?"时的输出结果。

```
#include<stdio.h>
void main()
{   char c;
    while((c=getchar())!='?') putchar(++c);
}
```

(8)
```
#include<stdio.h>
void main()
{   int i;
    for(i=1;i<6;i++)
    {   if(i%2){ printf("#"); continue; }
```

```
        printf(" * ");
    }
    printf("\n");
}
```

(9)

```
#include<stdio.h>
void main()
{   int s=0,i;
    for(i=1;;i++)
    {  if(s>50) break;
       if(i%2==0) s+=i;
    }
    printf("i=%d,s=%d\n",i,s);
}
```

(10)

```
#include<stdio.h>
void main()
{   int i,j,x=0;
    for(i=0;i<2;i++)
    {  x++;
       for(j=0;j<3;j++)
       {  if(j%2) continue;
          x++;
       }
       x++;
    }
    printf("=%d\n",x);
}
```

(11)

```
#include<stdio.h>
void main()
{   int j;
    for(j=1;j<=5;j++)
    {  if(j%2==0) printf(" * ");
       else continue;
       printf("#");
    }
    printf("$ \n");
}
```

(12)

```
#include<stdio.h>
```

```
void main()
{   int a,b;
    for(a=1,b=1;a<=100;a++)
    {   if(b==20) break;
        if(b%3==1){  b=3;  continue;  }
        b+=5;
    }
    printf("a=%d  b=%d\n",a,b);
}
```

(13)

```
#include<stdio.h>
void main()
{   int i=1,s=3;
    do
    {   s+=i++;
        if(s%7==0)continue;
        else ++i;
    } while(s<15);
    printf("%d\n",i);
}
```

2. 填空题

请根据题意在下面各程序中划线处填写适当的语句或表达式，使之能够运行并获得正确的结果。

(1) 求出 100 以内可被 13 整除的最大数。

```
#include<stdio.h>
void main()
{   int n;
    for(_____;_____;n--)
        if(n%13==0)   break;
    printf("the max is %d\n",n);
}
```

(2) 输入一个正整数后求出它的各位数之和并输出。例如若输入 123,则将各位之和 6 (即 1+2+3)输出。

```
#include<stdio.h>
void main()
{   int n,s=0;
    scanf("%d",&n);
    do
    {   s+=_____;
        n/=10;
    } while(n);
```

```
        printf("%d\n",s);
}
```

(3) 下列程序用于将从键盘输入的整数逆序输出。

```
#include<stdio.h>
void main()
{   int n1,n2;
    scanf("%d",&n2);
    while(_____)
    {   n1=n2%10;
        n2=_____;
        printf("%d",n1);
    }
}
```

(4) 下面的程序输出 3 到 1000 之间的所有素数,且每 5 个一行。

```
#include<stdio.h>
void main()
{   int i,j;
    int b=0,c=0;
    for(i=3;i<=1000;i++)
    {   for(j=2;j<=i-1;j++)
            if _____  {b=1; break;}
        if(!b)
        {   c++; printf("%4d",i);
            if _____ printf("\n");
        }
    }
}
```

(5) 从键盘上输入 10 个数,求其平均值。

```
#include<stdio.h>
void main()
{   int i;
    float f,sum;
    for(i=1,sum=0.0;i<11;i++)
    {   _____;
        _____;
    }
    printf("average=%f\n",sum/10);
}
```

(6) 以下程序的功能是:从键盘输入若干个学生的成绩,统计并输出最高成绩和最低成绩,当输入负数时结束输入。

```
#include<stdio.h>
```

```
void main()
{   float x,amax,amin;
    scanf("%f",&x);
    amax=x; amin=x;
    while(_____)
    {   if(x>amax) amax=x;
        if(_____) amin=x;
        scanf("%f",&x);
    }
    printf("\namax=%f\namin=%f\n",amax,amin);
}
```

(7) 下面程序的功能是：将从键盘输入的一组字符中统计出大写字母的个数 m 和小写字母的个数 n，并输出 m、n 中的较大者。

```
#include<stdio.h>
void main()
{   int m=0,n=0;
    char c;
    while((_____)!='\n')
    {   if(c>='A'&&c<='Z')   m++;
        if(c>='a'&&c<='z')   n++;
    }
    printf("%d\n",m<n?_____);
}
```

(8) 下面程序的功能是从 3 个红球、5 个白球、6 个黑球中任意取出 8 个球，且其中必须有白球，输出所有可能的方案。

```
#include<stdio.h>
void main()
{   int i,j,k;
    printf("红球    白球    黑球\n");
    for(i=0;i<=3;i++)
      for(_____;j<=5;j++)
      {   k=8-i-j;
          if(_____)printf("%3d%3d%3d\n",i,j,k);
      }
}
```

3. 改错题

下列各程序中存在两处或两处以上的错误，请仔细阅读程序，根据题意进行修改，使程序能够正确运行。

(1) 求 1～100 之间所有能被 7 整除的数之和。

```
#include<stdio.h>
void main()
```

```
{   int i,sum;
    for(i=1;i<100;i++)
       if(i%7==0) sum=sum+i;
    printf("the sum of 1-100 is %d\n",sum);
}
```

(2) 求斐波那契数列的前 40 项,每行输出 6 个数。这个数列有如下特点:第 1、2 两个数为 1、1,从第 3 个数开始,该数是其前面两个数之和。

```
#include<stdio.h>
void main()
{   int f1,f2;
    int i;
    f1=1;f2=1;
    for(i=1;i<=40;i++)
    {   printf("%15d%15d",f1,f2);
        if(i%6==0) printf("\n");
        f1=f1+f2;
        f2=f2+f1;
    }
}
```

(3) 从键盘输入若干字符,分别统计其中字母 a、b 和 c 的个数。

```
#include<stdio.h>
void main()
{   char ch;
    while((ch=getchar())!='\n')
    {   switch (ch)
        {   case 'a': a++;
            case 'b': b++;
            case 'c': c++;
        }
    }
    printf("the number of a、b and c is:%d %d %d",a,b,c);
}
```

(4) 输入某班级 10 名同学 5 门课程的成绩,分别统计每个学生 5 门课程的平均成绩。

```
#include<stdio.h>
void main()
{   int i,j;
    float grade,sum,average;
    sum=0;
    for(i=1;i<10;i++)
    {   for(j=1;j<5;j++)
        {   scanf("%f",&grade);
            sum=sum+grade;
```

```
            }
            average=sum/5;
            printf("no.%d average =%5.2f\n",i,average);
    }
}
```

(5) 在歌手大奖赛中有若干裁判为歌手打分,计算歌手最后得分的方法是,去掉一个最高分和一个最低分,取剩余成绩的平均分。输入歌手的成绩时,以－1作为输入结束标记。

```
#include<stdio.h>
void main()
{   float score,min,max,sum;
    int i=0;
    while(score!=-1)
    {   sum=sum+score;
        scanf("%f",&score);
        if(score>max) max=score;
        else min=score;
        i++;
    }
    sum=sum-max-min;
    printf("final score is %6.2f",sum/(i-2));
}
```

4. 编程题

(1) 编程实现输入一行字符,分别统计其中英文字母、空格、数字和其他字符的个数。

(2) 将一个大偶数 $a(a \geqslant 6)$ 分解成两个素数之和的形式。例如,若输入10,则输出 $10=3+7$ 和 $10=5+5$。

(3) 计算 y 的值:$y=1+1/x+1/x^2+1/x^3+1/x^4+\cdots(x>1)$ 直到某一项得值 $f \leqslant 10^{-6}$ 时为止。

(4) 有一堆零件,总数在100～200个之间,如果以4个零件为一组进行分组,则多2个零件;如果以7个零件为一组进行分组,则多3个;如果以9个零件为一组进行分组,则多5个零件。求这堆零件的总数(提示:用穷举法求解。即零件总数 x 从100～200 循环试探,如果满足所有几个分组已知条件,则此时的 x 就是一个解)。

(5) 一种体育彩票采用整数1、2、3、…、36表示36种体育运动,一张彩票可选择7种运动。编写程序,选择一张彩票的号码,使得这张彩票的7个号码之和是105且相邻两个号码之差按顺序依次是1、2、3、4、5、6(即如果第一个号码是1,则后续号码应是2、4、7、11、16、22)。

(6) 猜数游戏。由计算机"想"一个数请人猜,如果人猜对了,则结束游戏,否则计算机给出提示,告诉人所猜的数是太大还是太小,直到人猜对为止。计算机记录人猜的次数,以此可以反映出猜数者"猜"的水平。

(7) 假设某高速公路的一个收费站的收费标准为:小型车15元/车·次、中型车35元/车·次、大型车50元/车·次、重型车70元/车·次。编写程序,循环显示如图5.1所示的列表,请用户选择车型,根据用户的选择

(1) 小型车
(2) 中型车
(3) 大型车
(4) 重型车
(5) 退出

图 5.1

输出应缴纳的费用,直到用户选择"退出",程序结束。

三、实验步骤

本实验要求学生重点掌握循环结构的程序设计。重点要抓住循环结构的两大要素：循环体和循环控制条件的设置。其中循环体就是要反复执行的操作；而循环控制条件是循环体执行的前提条件。

1. 题目：编写程序,求 $1+2+3+\cdots+200$。
2. 算法分析

求累加和是一个典型的循环结构。如果用变量 i 表示当前被加数,则 i 的初值应该设置为 1。除此之外,用变量 sum 表示和,且未求和之前,sum 的初值应该设置为 0。这样,把当前被加数 i 加入到变量 sum 中,以及求出下一个被加数 i 的这两步操作就是循环体；而被加数 i 被加入到 sum 的前提条件,即循环控制条件就是 $i\leqslant 200$。

3. 根据分析,写出代码如下：

```c
#include<stdio.h>
void main()
{   int i,sum;
    i=1; sum=0;
    while(i<=200)
    {   sum=sum+i;
        i++;
    }
    printf("1+2+3+…+200=%d\n",sum);
}
```

4. 上机调试

（1）编辑源程序：

将源程序代码输入到 Visual C++ 代码编辑窗口中,如图 5.2 所示。

（2）编译、连接：

单击菜单 Build→Build,进行编译和连接,如图 5.3 所示。编译、连接中若发现错误将在输出窗口中显示,如图 5.4 所示。观察输出窗口发现编译、连接无误,严重错误（errors）0,警告错误（warnings）0,可执行文件已经生成。

（3）运行程序：

单击菜单 Build→! Execute Cpp1.exe,或者使用键盘组合键 Ctrl+F5,也可以直接单击工具栏中的按钮❗运行程序。运行结果如图 5.5 所示,结果正确,按任意键返回编辑窗口。

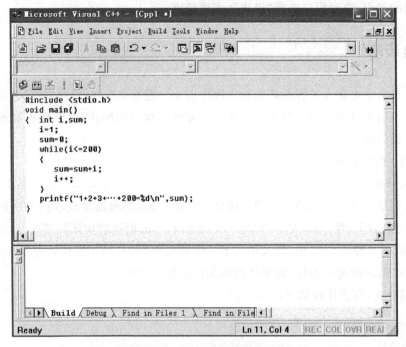

图 5.2 编辑源程序

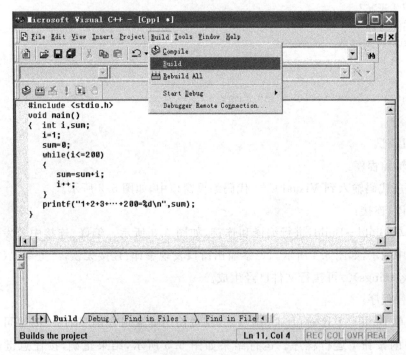

图 5.3 编译、连接程序

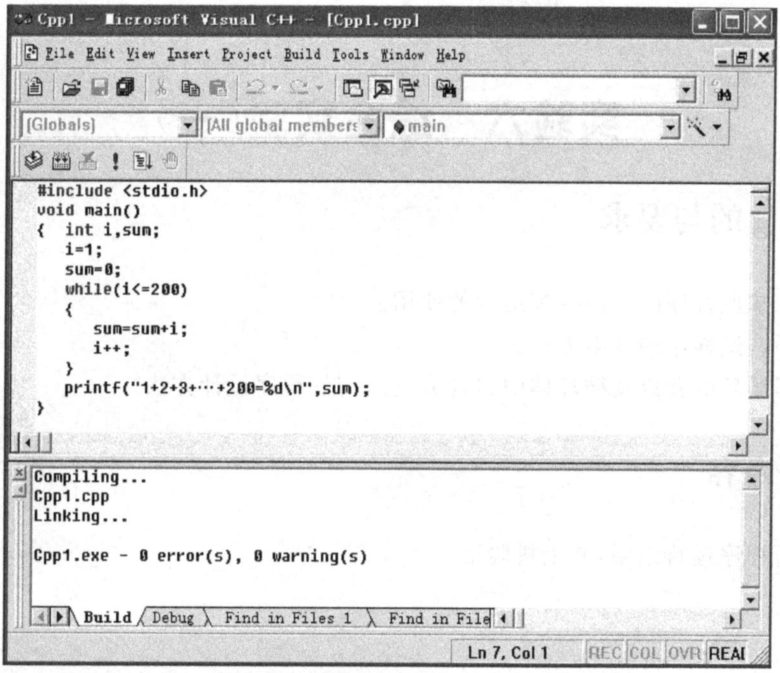

图 5.4 编译、连接成功信息窗口

图 5.5 运行程序窗口

四、思考题

1. while 语句和 do-while 语句的主要区别是什么？
2. 在什么情况下可能会造成死循环？应如何避免死循环？
3. break 语句和 continue 语句的区别是什么？用什么语句可以提前结束循环？
4. for(;;){…} 在形式上是合法的循环语句吗？

实验六 综合实验(一)

一、实验目的与要求

1. 熟练掌握程序的三种控制结构的使用。
2. 掌握调试程序的基本方法。
3. 掌握循环嵌套以及循环体中包含分支语句的程序设计方法。

二、实验内容

1. 分析程序运行结果,并上机验证。

(1)

```
#include<stdio.h>
void main()
{   int x=1,y=0,a=0,b=0;
    switch(x)
    {   case 1:
            switch (y)
            {   case 0: a++; break;
                case 1: b++; break;
            }
        case 2:
        {   a++;   b++;   break;   }
    }
    printf("a=%d,b=%d\n",a,b);
}
```

(2) 分析当输入 About 时下列程序的输出结果。

```
#include<stdio.h>
void main()
{   int i=0;
    char c;
    while ((c=getchar())!='\n')
    {   switch(c)
        {   case 'A':i++; break;
            case 'B':++i;
            default: putchar(c);i++;
        }
        putchar('*');
```

        }
}

(3)

```
#include<stdio.h>
void main()
{   int j;
    for(j=4;j>=2;j--)
    switch(j)
    {   case 0: printf("%4s","ABC");
        case 1: printf("%4s","DEF");
        case 2: printf("%4s","GHI");break;
        case 3: printf("%4s","JKL");
        default: printf("%4s","MNO");
    }
    printf("\n");
}
```

(4)

```
#include<stdio.h>
void main()
{   int a,b;
    for(a=1,b=1;a<=100;a++)
    {   if(b>=20) break;
        if(b%3==1){ b+=3;   continue; }
        b-=5;
    }
    printf("a=%d   b=%d\n",a,b);
}
```

(5)

```
#include<stdio.h>
void main()
{   int m,n,sign,t;
    scanf("%d%d",&m,&n);
    while(m*n)
    {   if(m>=0&&n>=0||m<=0&&n<=0) sign=0;
        else sign=1;
        m=m>0?m:-m; n=n>0?n:-n;
        t=0;
        while(n--) t+=m;
        printf("The result is:");
        if (sign) printf("-");
        printf("%d\n",t);
        scanf("%d%d",&m,&n);
```

    }
}

2. 填空题

请根据题意在下面各程序中划线处填写适当的语句或表达式,使之能够运行并获得正确的结果。

(1) 以下程序的功能是:输出 100 以内能被 3 整除且个位数为 6 的所有整数。

```
#include<stdio.h>
void  main()
{   int i,j;
    for(i=0; _____ ;i++)
    {  j=i*10+6;
       if(          )continue;
       printf("%d ",j);
    }
}
```

(2) 输出如图 6.1 所示的图形。

```
#include<stdio.h>
void main()
{  int i,j;
   for(i=1;i<=_____;i++)
   {  for(j=_____;j<=9;j++)printf("%2d",j);
      printf("\n");
   }
}
```

```
123456789
23456789
3456789
456789
56789
6789
789
89
9
```

图 6.1 数字图形

(3) 从键盘输入的字符中统计数字字符的个数,用换行符结束循环。

```
#include<stdio.h>
void main()
{   int n=0,c;
    c=getchar();
    while(_____)
    {  if(_____)n++;
       c=getchar();
    }
    printf("%d\n",n);
}
```

(4) 用"辗转相除法"求两个正整数的最大公约数。辗转相除法的思路是:将两数中较小数作为除数,较大数作为被除数,两数相除所得的余数若为 0,则此时的除数即为最大公约数。若余数不为 0,则原来的除数作为新的被除数,余数作为新的除数,继续求余,直到余数为 0。

#include<stdio.h>

```
void  main()
{  int r,m,n;
   scanf("%d%d",&m,&n);
   if(m<n)_____;
   r=m%n;
   while(r){ m=n; n=r; r=_____;}
   printf("%d\n",n);
}
```

(5) 计算某年某月有几天。其中判别闰年的条件是：能被 4 整除但不能被 100 整除的年是闰年，能被 400 整除的年也是闰年。

```
#include<stdio.h>
void main()
{  int yy,mm,len;
   printf("year,month=");
   scanf("%d%d", &yy,&mm);
   switch(mm)
   { case 1:case 3:   case 5:   case 7:   case 8:   case10:
     case12:_____;  break;
     case 4:case 6:   case 9:
     case 11:len=30;  break;
     case 2:if(yy%4==0&&yy%100!=0||yy%400)_____;
            else _____;
            break;
     default:printf("input error");  break;
   }
   printf("the length of %d%d is %d\n",yy,mm,len);
}
```

(6) 输出 1～100 之间每位数的乘积大于每位数的和的数。例如 89 每位数的乘积为 8*9=72,每位数的和是 8+9=17。因此 89 每位数的乘积大于每位数的和。

```
#include<stdio.h>
void main()
{  int n,k=1,s=0,m;
   for(n=1;n<=100;n++)
   {  k=1;s=0;
      _____;
      while(_____)
      {  k*=m%10;
         s+=m%10;
         _____;
      }
      if(k>s) printf("%d",n);
   }
}
```

(7) 输出如图 6.2 所示方阵。

```
#include<stdio.h>
void main()
{  int i,j,k=0;
   for(i=1;i<=4;i++)
   {  for(j=1;j<=4;j++)
      {  _____;
         _____;
      }
      printf("\n");
   }
}
```

```
1   2   3   4
5   6   7   8
9  10  11  12
13 14  15  16
```
图 6.2 方阵

3. 编程题

(1) 有一分数序列 2/1,3/2,5/3,8/5,13/8,…,求出这个数列的前 20 项之和。

(2) 输出 100 到 999 之间所有整数中各位数字之和为 x(x 从键盘输入)的整数。

(3) 任意输入两个日期,计算这两个日期之间的天数。

(4) 求 1!+2!+…+20!。

(5) 求解爱因斯坦数学题:有一条长阶梯,若每步跨 2 阶,则最后剩余 1 阶;若每步跨 3 阶,则最后剩余 2 阶;若每步跨 5 阶,则最后剩余 4 阶;若每步跨 6 阶,则最后剩余 5 阶;若每步跨 7 阶,则最后才正好一步不剩。请问,这条阶梯共有多少阶?

(6) 每个苹果 0.8 元,第一天买 2 个苹果,第二天开始,每天买前一天的 2 倍,直至购买的苹果个数达到不超过 100 的最大值,求每天平均花多少钱?

(7) 100 匹马驮 100 担货,大马一匹驮 3 担,中马一匹驮 2 担,小马一匹驮 1 担。计算大、中、小马的数目。

(8) 编写输出如图 6.3 所示图案。

(9) 编写程序,模拟石头、剪子、布游戏。规则是:石头砸剪子,剪子剪布,布包石头。程序先随机生成石头、剪子和布中的一个,然后询问用户选择石头、剪子和布中的哪一个,并判断输赢,接着,询问用户是继续游戏还是结束游戏,如果用户选择继续,则重复刚才的操作,否则,程序结束。要求在程序结束前输出用户与计算机的输赢比例。

图 6.3

三、实验步骤

1. 题目

利用循环输出如图 6.4 所示图形。

2. 算法分析

在一行上输出 n 个星号,显然是一个循环问题。其中,循环体就是输出星号的操作,而循环条件就是 i≤n,其中,i 为循环控制变量,且 i 的初值为 1。具体如下:

图 6.4

```
for(i=1;i<=n;i++)
    printf("*");
```

当需要输出 m 行星号时,很明显就是一个嵌套的循环。其中外层循环就是用来控制多行信息的输入,而内层则是输出一行星号。具体框架为:

```
For(j=1;j<=m;j++)
    输出一行信息;
```

3. 根据分析,写出代码如下:

```
#include<stdio.h>
void main()
{   int i,j;
    for(j=1;j<=5;j++)
    {   for(i=1;i<=6-j;i++)
            printf("*");
        printf("\n");
    }
}
```

4. 上机调试

将源程序代码输入到 Visual C++ 代码编辑窗口中,进行编译和连接,排查并纠正其中可能存在的语法错误,无误后运行程序,程序执行结果如图 6.5 所示,正确。有时程序虽然没有语法错误,但是程序执行后输出的结果与题意不符,说明程序中使用的算法存在问题,需要根据题意修改完善,以满足题意要求。

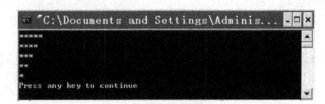

图 6.5　程序运行结果

实验七 一维数组与二维数组

一、实验目的与要求

1. 掌握一维数组和二维数组的定义、赋值、数组元素的引用和数组的输入输出方法。
2. 掌握与数组有关的数值计算方法,如矩阵运算等。
3. 掌握数组定义方法、数组元素的引用方法以及程序设计的相关算法。

二、实验内容

1. 填空题

请根据题意在下面各程序中划线处填写适当的语句或表达式,使之能够运行并获得正确的结果。

(1) 从键盘输入 20 个数,并以每行 5 个数据的形式输出 a 数组。

```
#include<stdio.h>
#define N 20
void main()
{   int a[N],i;
    for(i=0;i<N;i++) scanf("%d",_____);
    for(i=0;i<N;i++)
    {   if(_____)printf("\n");
        printf("%d ",a[i]);
    }
    printf("\n");
}
```

(2) 输入 20 个数,输出它们的平均值,并且输出与平均值之差的绝对值最小的数组元素。

```
#include<stdio.h>
#define N 20
_____
void main()
{   float a[N],average=0,s,t=a[0];
    int i,k;
    for(i=0;i<N;i++)
    scanf("%f",&a[i]);
    for(i=0;i<N;i++)
    average+=a[i];
```

```
        _____;
    s=fabs(a[0]-average);
    for(i=1;i<N;i++)
        if(fabs(a[i]-average)<s)
        { s=fabs(a[i]-average); t=a[i]; }
    printf("%d个数的平均值为%.1f,与平均值之差的绝对值为最小的数组元素值为%.1f",N,
    average,t);
}
```

(3) 以下程序是输出数组 s 中最小元素的下标。

```
#include<stdio.h>
void main()
{   int k,p;
    int s[]={1,-18,7,2,-20,5};
    for(p=0,k=p;p<6;p++) if(s[p]<s[k]) _____;
    printf("最小元素的下标是%d\n",k);
}
```

(4) 以下程序给偶数行的方阵中所有边上的元素和两对角线上的元素置 1,其他元素置 0。要求对每个元素只置一次值。最后按矩阵形式输出。

```
#include<stdio.h>
void main()
{   int a[10][10],i,j;
    for(i=0;i<10;i++)
    {   for(j=0;j<10;j++)
        {   if(_____)                    /*边界元素置1*/
                _____;
            else                            /*非边界元素置0*/
                _____;
        }
    }
    for(i=0;i<10;i++)
    {   for(j=0;j<10;j++) printf("%2d",a[i][j]);
        printf("\n");
    }
}
```

(5) a 是一个 3×3 的矩阵,输入 a 的元素,如果 a 是下三角矩阵,输入 YES,否则,输出 NO(下三角矩阵,即主对角线以上的元素都为 0,主对角线为从矩阵的左上角到右下角的连线)。

```
#include<stdio.h>
void main()
{   int flag,i,k,; int a[3][3];
    for(i=0;i<3;i++)
```

```
    for(k=0;k<3;k++)
       scanf("%d",&a[i][k]);
    _____;
    for(i=0;i<3&&flag;k++)
       for(k=i+1;k<3&&flag;k++)
          if(a[i][k]!=0) _____;
    if(flag) printf(YES\n);
    else printf("NO\n");
}
```

(6) 数组 a 包括 10 个整型元素。以下程序的功能是求出 a 中各相邻两个元素的和，并将这些和存在数组 b 中，按每行 3 个元素的形式输出。

```
#include<stdio.h>
void main()
{  int a[10],b[10],i;
   for(i=0;i<10;i++)
       scanf("%d",&a[i]);
   for(i=1;i<10;i++)
       _____
   for(i=1;i<10;i++)
   {  printf("%3d",b[i]);
      if(_____)  printf("\n");
   }
}
```

(7) 以下程序是将矩阵 a、b 的乘积存入矩阵 c 中并按矩阵形式输出。

```
#include<stdio.h>
void main()
{  int a[3][2]={3,-2,-5,1,4,2};
   int b[2][2]={8,-7,-6,11};
   int i,j,k,s,c[3][2];
   for(i=0;i<3;i++)
       for(j=0;j<2;j++)
       {  for(_____;k<2;k++)
             s+=_____
          c[i][j]=s;
       }
   for(i=0;i<3;i++)
   {  for(j=0;j<2;j++)
         printf("%6d",c[i][j]);
      _____
   }
}
```

2. 改错题

下列各个程序中/****************/的下一行中有错误,请仔细阅读程序,并根据题意改正。

(1) 程序功能:给数组 a 输入数据并以每行 4 个数据的形式输出。

```
#include<stdio.h>
void main()
{   int a[20],i;
    for(i=0;i<20;i++)
    /************************/
        scanf("%d",a[i]);
    for(i=0;i<20;i++)
    {   /************************/
        if(i%4=0)  printf("\n");
        printf("%3d",a[i]);
    }
}
```

(2) 程序功能:求矩阵 a 的两条对角线上的元素之和。

```
#include<stdio.h>
void main()
{   int a[3][3]={1,3,6,2,4,8,11,14,15},s1=0,s2=0,i,j;
    /************************/
    for(i=0;i<=3;i++)
        for(j=0;j<3;j++)
            if(i==j) s1=s1+a[i][j];
    for(i=0;i<3;i++)
        /************************/
        for(j=2;j>=0;j++)
            if((i+j)==2)s2=s2+a[i][j];
    printf("s1=%d,s2=%d\n",s1,s2);
}
```

(3) 程序功能:求 1000 以内的水仙花数(提示:水仙花数是指一个 3 位正整数,其各位数字的立方之和等于该正整数。例如:407=4*4*4+0*0*0+7*7*7,所以 407 是一个水仙花数)。

```
#include<stdio.h>
void main()
{   int x,y,z,a[8],m,i=0;
    printf("水仙花数是:\n");
    /************************/
    for(m=100;m<1000;m++);
    {   x=m/100;
        y=m/10-x*10;
        z=m%10;
```

```
       /***************************/
       if(x*100+y*10+z=x*x*x+y*y*y+z*z*z)
       {
          a[i]=m;
          i++;
       }
   }
   for(x=0;x<i;x++)
      printf("%6d",a[x]);
}
```

(4) 程序功能：将十进制数转化为八进制数。

```
#include<stdio.h>
void main()
{  int i=0,n,j,num[20];
   printf("请输入要转换的数据：\n");
   /***************************/
   scanf("%d",n);
   do
   {  i++;
      num[i]=n%8;
      n=n/8;
      /*********************/
   }while(n!=0)
   for(j=i;j>=1;j--)  printf("%d",num[j]);
}
```

(5) 程序功能：检查一个二维数组是否对称（即对所有 i、j 都有 a[i][j]＝a[j][i]）。

```
#include<stdio.h>
void main()
{  int a[4][4]={2,3,4,5,3,3,6,7,4,6,4,8,5,7,8,4};
   int i,j,found=0;
   for(j=0;j<4;j++)
      for(i=j+1;i<4;i++)
          /*********************/
          if(a[j][i]==a[i][j])
          {found=1;break;}
   /**************************/
   if(found=1)
       printf("No");
   else printf("Yes");
}
```

(6) 程序功能：将矩阵 a、b 的和存入矩阵 c 中并按矩阵形式输出。

```
#include<stdio.h>
```

```
void main()
{   int a[3][4]={{4,-3,7,5},{1,0,4,-3},{7,9,1,2}};
    int b[3][4]={{-3,1,2,4},{5,-2,8,7},{7,9,1,2}};
    /**********************/
    int i,j;c[3][4];
    for(i=0;i<3;i++)
        /************************/
        for(j=0;j<=4;j++)
            c[i][j]=a[i][j]+b[i][j];
    for(i=0;i<3;i++)
    {   for(j=0;j<4;j++)
            printf("%3d",c[i][j]);
        printf("\n");
    }
}
```

（7）设数组 a 中的元素均为正整数，以下程序是求 a 中偶数的个数和偶数的平均值。

```
#include<stdio.h>
void main()
{   int a[10]={2,3,4,5,6,7,8,9,10,11};
    int k,s,i;
    float ave;
    for(k=s=i=0;i<10;i++)
    {   /**************************/
        if(a[i]%2!=0) break;
        s+=a[i];
        k++;
    }
    /*************************/
    if(k==0)
    {ave=s/k;printf("%d,%f\n",k,ave);}
}
```

（8）该程序功能是：按以下形式构成一个杨辉三角形，要求输出 10 行。杨辉三角形各行是 $(a+b)^n$ 展开后各项的系数（n=0,1,2,3,…）。输出杨辉三角实际上是输出 10×10 方阵的左下半三角。

```
1
1   1
1   2   1
1   3   3   1
1   4   6   4   1
1   5  10  10   5   1
……
#include<stdio.h>
#define N 11
```

```
void main()
{  int i,j,a[N][N];
   for(i=1;i<N;i++)
   {  a[i][1]=1;
      a[i][i]=1;
   }
   /***************/
   for(i=1;i<N;i++)
   /***************/
     for(j=2;j<=i;j++)
        a[i][j]=a[i-1][j-1]+a[i-1][j];
   for(i=1;i<N;i++)
   {  /***************/
      for(j=1;j<N;j++)
        printf("%5d",a[i][j]);
      printf("\n");
   }
}
```

3. 编程题

(1) 编程,输入 1 个正整数 n(1<n≤10),再输入 n 个整数,按绝对值从小到大排序后输出。输入/输出示例:

输入整数个数:10
输入 10 个整数:-11 2 8 5 -3 -16 9 7 6 10
排序后:2 -3 5 6 7 8 9 10 -11 -16

(2) 输入两个数组(数组元素自定),输出在两个数组中不同时出现的元素。

(3) 输入一个 n 行 m 列(n≤4,m≤4)的数组,先以 n 行 m 列的格式输出该数组,然后找出该数组中值最小的元素,输出该元素及其行下标和列下标。

(4) 计算多项式 $a_0+a_1*x+a_2*x*x+a_3*x*x*x+\cdots$ 的值,并将其值以格式"%f"输出。

(5) 编写程序,生成并打印 Fabonaci 的前 20 项,该数列第 1、2 项分别为 1 和 1,以后每项等于前两项之和。要求生成的 20 个数存在一维数组 x 中,并按每行 4 个数的形式输出。

(6) 编写程序,从键盘上输入单精度型一维数组 a[10],计算并输出 a 数组中所有元素的平均值。

(7) 输入一个 3×5 的整数矩阵,输出其中最大值、最小值和它们的下标。

(8) 某班 30 名学生的三科成绩表如下:

课程一 课程二 课程三
  ⋮   ⋮   ⋮

试编写程序,输入这 30 名学生的三科成绩,计算并输出每科成绩的平均分。

(9) 编程使 N*N 矩阵第一列与最后一列对调、第二列与倒数第二列对调……,其他依次类推。

(10) 输入一个整型数据,输出每位数字,其间用逗号分隔。例如,输入为:12345;输出为：1,2,3,4,5。

(11) 求数组 a 的 10 个数的平均值 v，将大于等于 v 的数组元素进行求和，并将结果以格式"%.4f"输出。

(12) 数组 a 的长度为 m+n，编写程序把数组的前 m 个元素和后 n 个元素交换。

三、实验步骤

本实验要求掌握一维、二维数值型数组的应用，举例说明如何利用数组处理一组数据。

1. 题目：输入螺旋方阵的阶数 n，编程生成 n×n 的螺旋方阵。例如输入 5，生成如下矩阵：

```
 1   2   3   4   5
16  17  18  19   6
15  24  25  20   7
14  23  22  21   8
13  12  11  10   9
```

2. 算法分析

定义一个二维数组 a，设 i、j 分别代表矩阵行、列坐标，变量 k 从 1 到 n^2 依次存入数组 a[i][j]中，要确定 i,j 的变化情况，保证 k 的值赋值到正确的数组坐标。首先根据方阵的阶数确定螺旋的圈数，按照圈数做循环，根据螺旋方向，依次确定螺旋顶部的值，从左到右，行不变列变，k 的值每次增 1；确定螺旋右边的值，从上到下，行变列不变；确定螺旋底部的值，从右到左，行不变列变；确定螺旋左边的值，从下到上，行变列不变。将方阵每一个元素的值都确定完毕后，通过二重循环输出该二维数组即可。

3. 根据以上分析，编写如下程序代码：

```c
#include<stdio.h>
void main()
{   int i,j,n,k=1,a[50][50];
    printf("请输入螺旋方阵的阶数：");
    scanf("%d",&n);
    for(i=0;i<(n+1)/2;i++)              /*i 控制生成方阵的圈数*/
    {                                    /*j 控制生成一条边的数据*/
        for(j=i;j<n-i;j++)              /*顶边，从左到右,行不变列变*/
            a[i][j]=k++;                /*k 从 1 递增到 n*n*/
        for(j=i+1;j<n-i;j++)            /*右边，从上到下,行变列不变*/
            a[j][n-i-1]=k++;
        for(j=n-i-2;j>=i;j--)           /*底边，从右到左,行不变列变*/
            a[n-i-1][j]=k++;
        for(j=n-i-2;j>=i+1;j--)         /*左边，从下到上,行变列不变*/
            a[j][i]=k++;
    }
    for(i=0;i<n;i++)                     /*输出二维数组*/
    {   for(j=0;j<n;j++)
```

```
            printf("%5d",a[i][j]);
        printf("\n");
    }
}
```

4. 上机调试

将源程序代码录入到 Visual C++ 代码编辑窗口中,接着进行编译和连接,排查并纠正其中可能存在的语法错误,无误后运行程序。本程序运行结果如图 7.1 所示。

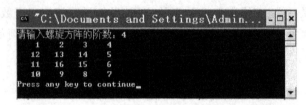

图 7.1　程序运行结果

四、思考题

1. 定义数组时应注意哪些问题？引用数组元素时应注意哪些问题？
2. 数组中下标有何作用？对数组元素的操作是通过什么来实现的？
3. 用 for 语句、while 语句和 do…while 语句都可对数组元素赋值,请问在循环控制变量的处理上有何异同？
4. 如何理解"二维数组的数组元素是一个一维数组"这句话？
5. 如果有以下语句：int i,j,a[3][4];则 a、a[i]各表示什么？

实验八 字 符 串

一、实验目的与要求

1. 熟练掌握字符型数组的定义、赋值和输入输出的方法。
2. 熟练掌握用字符型数组来存储和处理字符串常量的方法。
3. 掌握字符串处理函数的使用。
4. 掌握字符数组和数值型数组在处理上的区别和联系。
5. 复习字符常量和字符串常量的概念。
6. 掌握字符型数组和字符串函数在实际编程中的应用。

二、实验内容

1. 填空题

请根据题意在下面各程序中划线处填写适当的语句或表达式,使之能够运行并获得正确的结果。

(1) 编写程序,功能是统计一个长度为 2 的字符串在另一个字符串中出现的次数。例如,输入的字符串为:asd asasdfg asd as zx67 asd mklo,子字符串为 as,则应输出 6。

```
#include<stdio.h>
#include<string.h>
void main()
{   char str[81],substr[3];
    int i,n=0;
    printf("请输入字符串:");
    gets(str);
    printf("请输入 2 个字符的字符串:");
    _____;
    puts(str);
    puts(substr);
    for(i=0;i<strlen(str);i++)
        if(_____)   n++;
}
```

(2) 以下程序是将一个字符串 str 的内容颠倒过来。如输入 abcde,输出 edcba。

```
#include<stdio.h>
_____
void main()
{   int i,j,k;
```

```
    char str[80];
    printf("请输入一个字符串(少于80个字符):");
    gets(str);
    for(i=0,j=strlen(str)-1;_____;i++,j--)        /*头和尾交换,直到中间*/
    {   k=str[i];
        str[i]=str[j];
        str[j]=k;
    }
    printf("颠倒后的字符串:%s\n",str);
}
```

(3) 从键盘输入一串字符,下面程序能统计输入字符中各个大写字母的个数。用#号结束输入。该程序利用了字符的 ASCII 码和整数的对应方法,数组 c 的下标为 0 到 25,当输入为 ca='A',则 ca－65＝0,c[0]的值加 1,依次类推。而输出时 c[i]的下标 i+65 正好又是相应的字母。

```
#include<stdio.h>
void main()
{   int c[26],i; char ca;
    for(i=0;i<26;i++)   c[i]=_____;
    printf("请输入一个以#结束的大写字母串:\n");
    scanf("%c",&a);
    while(_____)
    {   if((ca>='A')&&(ca<='Z')) c[ca-65]+=1;
        _____;
    }
    for(i=0;i<26;i++)
        if(c[i]) printf("字符%c有%d个\n",i+65,c[i]);
}
```

(4) 分别统计字符串中英文字母、数字和其他字符出现的次数。

```
#include<strio.h>
#include<ctype.h>
void main()
{   char a[80]; int n[3]={0},i;
    char xx[3][18]={"英文字母个数","数字字符个数","其他字符个数"};
    printf("请输入一个字符串:\n");
    gets(a);
    for(i=0;_____;i++)
    {   if(_____)  n[0]++;
        else if(_____)  n[1]++;
        else n[2]++;
    }
    for(i=0;i<3;i++)
    {   printf("%s",xx[i]);
        printf("%d\n",n[i]);
```

（5）以下程序是从键盘上输入一个字符串，将组成字符串的所有非英文字母的字符删除后输出。例如，输入 chi 56, nes －78e, 输出 chinese。

```
#include<stdio.h>
#include<ctype.h>
void main()
{   char str[256];
    int i,j,k=0,n;
    printf("请输入一个字符串：");
    gets(str);
    n=strlen(str);
    for(i=0;i<n;i++)
        if(_____)
        {   _____;
            _____;
        }
    str[k]='\0';
    printf("去除非英文字符后的字符串：");
    printf("%s\n",str);
}
```

（6）以下程序是分别统计字符串中所有英文字母中的各元音字母个数。

```
#include<stdio.h>
#include<ctype.h>
void main()
{   char a[81]; int n[5]={0},i;
    gets(a);
    for(i=0;_____;i++)
        switch(_____)
        {   case 'a':n[0]++;break;
            case 'e':n[1]++;break;
            case 'i':n[2]++;break;
            case 'o':n[3]++;break;
            case 'u':n[4]++;
        }
    printf("元音 a 的个数为%d,元音 e 的个数为%d",n[0],n[1]);
    printf("元音 i 的个数为%d,元音 o 的个数为%d,元音 u 的个数为%d",n[2],n[3],n[4]);
}
```

（7）以下程序是将两个字符串连接起来（不能使用库函数）。

```
#include<stdio.h>
#include<ctype.h>
void main()
```

```
{   char s1[80],s2[40];
    int i=0,j=0;
    printf("请输入第一个字符串:");
    scanf("%s",s1);
    printf("请输入第二个字符串:");
    scanf("%s",s2);
    while(s1[i]!='\0')
        _____;
    while(s2[j]!='\0')
        s1[i++]=s2[j++];
    _____;
    printf("\n连接后的字符串为:%s\n",s1);
}
```

(8) 以下程序是把字符串中所有的字母改写成该字母的下一个字母,最后一个字母 z 改写成字母 a。大写字母仍为大写字母,小写字母仍为小写字母,其他的字符不变。例如,输入的字符串为 Hello.456,程序运行后输出 Ifmmp.456。

```
#include<string.h>
#include<stdio.h>
#define N 81
void main()
{   char s[N];
    int i,n;
    printf("请输入一个字符串:");
    gets(s);
    n=strlen(s);
    for(i=0;i<n;i++)
    {   if(_____)_____;
        else if(_____) s[i]='a';
        else if(s[i]=='Z') s[i]='A';
    }
    printf("变换后的字符串:");
    puts(a);
}
```

(9) 以下程序是输入 5 个字符串,输出其中最小的字符串。

```
#include<stdio.h>
_____
void main()
{   int i;
    char sx[80],smin[80];
    scanf("%s",sx);
    strcpy(smin,sx);
    for(i=1;i<5;i++)
    {
```

```
        scanf("%s",sx);
        if(_____) strcpy(smin,sx);
    }
}
```

2. 改错题

下列各个程序中"/***************/"的下一行中有错误,请仔细阅读程序,并根据题意改正。

(1) 程序功能:判断用户输入的任意一个字符串是否为"回文串"。所谓"回文串"是指从开头读和从末尾读均为相同的字符串,例如"rotator"就是回文串。

```
#include<stdio.h>
#define N 50
void main()
{   char a[N];
    int i=0,num=0,flag=0;
    /***************/
    scanf("%c",a);
    do{
        num++;
        /***************/
    }while(a[i]!='\0');
    do{  /***************/
        if(a[i]!=a[num-i])
        {   flag=1; break;  }
        i++;
    } while(i<num);
    if(flag==1)  printf("此数组不是回文串\n");
    else  printf("此数组是回文串\n");
}
```

提示:关键要弄清楚程序中变量所代表的含义:变量 flag 作为回文串的标志,当 flag 为 0 时此字符串为回文串,当 flag 为 1 时此字符串不是回文串;变量 num 用来存放字符串的真实长度。当从键盘上输入字符串赋给字符数组后,则字符串为已知字符串,对已知字符串处理往往用"\0"作为字符串处理结束的标志。

(2) 程序功能:将字符串 a 中下标值为偶数的元素由小到大排序,其他元素不变。

```
#include<stdio.h>
void main()
{   char a[]="labchmdes",t;
    int i,j;
    for(i=0;i<7;i+=2)
        /*********************/
        for(j=i+2;j<=9; j++)
            /*********************/
            if(a[i]<a[j])
```

```
            {t=a[i];a[i]=a[j];a[j]=t;j++;}
    puts(a);
}
```

(3) 下面程序的功能是在每个字符串中找出最大字符,并按一一对应的顺序放入一维数组 a 中,即第 i 个字符串中的最大字符放入 a[i]中,输出每个字符串中的最大字符。

```
#include<stdio.h>
void main()
{   char s[10][20];
    /***********************/
    int a[10];i,j;
    for(i=0;i<10;i++)
        gets(s[i]);
    for(i=0;i<10;i++)
    {   a[i]=s[i][0];
        for(j=1;s[i][j]!='\0';j++)
            /***********************/
            if(a[i]>s[i][j]) a[i]=s[i][j];
    }
    for(i=0;i<10;i++) printf("%d  %c",i,a[i]);
}
```

(4) 在任意的字符串 a 中,将与字符 c 相等的所有元素的下标值分别存放在整型数组 b 中。

```
#include<stdio.h>
void main()
{   char a[80];
    /*************************/
    int i,b[80];k=0;
    gets(a);
    for(i=0;a[i]!='\0';i++)
        /**************************/
        if(a[i]='c')
        {   b[k]=i;
            k++;
        }
    for(i=0;i<k;i++)  printf("%3d",b[i]);
}
```

(5) 程序功能:在 3 个字符串中找出最小的,并将最小的字符串输出。

```
#include<stdio.h>
#include<string.h>
void main()
{   char s[20],str[3][20];
    int i;
```

```
/***********************/
for(i=0;i<=3;i++)
    gets(str[i]);
strcpy(s,(strcmp(str[0],str[1])<0?str[0]:str[1]));
/***********************/
if(strcmp(str[2],s)>0)strcpy(s,str[2]);
printf("%s\n",s);
}
```

(6) 从键盘输入一行字符,统计其中有多少个单词,单词之间用空格分隔。

```
#include<stdio.h>
void main()
{   char s[80],c1,c2='';
    int i=0,num=0;
    gets(s);
    /***********************/
    while(s[i]!='\0');
    {   c1=s[i];
        if(i==0)c2=' ';
        else c2=s[i-1];
        /***********************/
        if(c1!=' '||c2==' ')num++;
        i++;
    }
    printf("总共有%d个单词.\n",num);
}
```

(7) 程序功能:将字符串 s 中所有的字符 c 删除。

```
#include<stdio.h>
void main()
{   char s[80];
    int i,j;
    /****************/
    scanf("%s",&s);
    for(i=j=0;s[i]!='\0';i++)
        if(s[i]!='c')
        {   /****************/
            s[i]=s[j];
            j++;
        }
    s[j]='\0';
    puts(s);
}
```

(8) 有已排好序的字符串 a,现将字符串 s 中的每个字符按 a 中元素的规律插入到 a 中。

```c
#include<stdio.h>
#include<string.h>
void main()
{   char a[20]="aehjkmosw";
    char s[]="fcna";
    int i,k,j;
    for(k=0;s[k]!='\0';k++)
    {   j=0;
        /*************************/
        while(s[k]>=a[j]||a[j]!='\0')
            j++;
        for(i=strlen(a);i>=j;i--)
            /*************************/
            a[i]=a[i+1];
        a[j]=s[k];
    }
    puts(a);
}
```

3. 编程题

（1）编写程序，输入一个字符串，将其中所有的大写字母加 3，所有的小写字母减 3，然后再输出加密后的字符串。

（2）输入一个字符串（少于 80 个字符），再输入一个字符，统计并输出该字符在字符串中出现的次数。

（3）有 3 行字符串，每行 80 个字符，要求分别统计出其中英文大写字母、小写字母、数字、空格以及其他字符的个数。

（4）将字符数组 s2 中的全部字符复制到字符数组 s1 中。要求：不使用库函数 strcpy。

（5）从键盘输入一个数字字符串，将其转换为整数后输出。

三、实验步骤

本实验要求掌握字符型数组的应用以及字符型数组与数值型数组在使用中的不同之处。举例说明利用字符型数组处理字符串的方法。

1. 题目：输入 5 个国家的名字，对名字按从小到大的顺序排序，输出排序后的结果。

2. 算法分析

可以用一个二维字符型数组来存储五个国家的名字，将这个二维数组看成是 5 个一维字符型数组，即对这 5 个字符串排序，可用冒泡排序法排序。在程序中定义一临时数组 temp 作为排序时交换用的中间变量。使用字符串比较函数 strcmp 实现字符串的排序。

3. 根据分析，写出代码如下：

```c
#include<stdio.h>
```

```
#include<string.h>
void main()
{   char name[5][20],temp[20];
    int i,j;
    printf("请输入五个国家的名字：\n");
    for(i=0;i<5;i++)
        gets(name[i]);
    for(i=0;i<4;i++)
        for(j=0;j<4-i;j++)
            if(strcmp(name[j],name[j+1])>0)
            {   strcpy(temp,name[j]);
                strcpy(name[j],name[j+1]);
                strcpy(name[j+1],temp);
            }
    printf("排序后的结果为：\n");
    for(i=0;i<5;i++)
        puts(name[i]);
}
```

4. 上机调试

将源程序代码录入到 Visual C++ 代码编辑窗口中，然后进行编译与连接，排查其中可能存在的语法错误并纠正，无误后运行程序，运行结果如图 8.1 所示。

图 8.1　程序运行结果

四、思考题

1. 什么是字符串变量？什么是字符常量?"a"与'a'有什么区别？

2. 如果定义字符型数组 s[100]，在输入时是否必须输入 100 个字符？输入字符的个数与数组的长度是否必须一致？如何控制字符的实际个数？

3. 对二维字符型数组 s[10][10]的输入输出，必须严格地按行和列的大小操作吗？两者有何不同？实际字符个数又是如何控制的？

实验九 函 数（一）

一、实验目的与要求

1. 理解函数的概念,掌握函数的定义和调用方法。
2. 了解函数的形参和实参之间的对应关系。
3. 了解函数的返回值的概念。
4. 掌握函数的递归调用和嵌套调用。

二、实验内容

1. 填空题

请根据题意在下面各程序中划线处填写适当的语句或表达式,使之能够运行并获得正确的结果。

(1) 已知长方形的长和宽,定义一个函数,求长方形的面积。

```
#include<stdio.h>
void main()
{    _____;
     float x,y,s;
     printf("请输入两个数：");
     scanf("%f %f",&x,&y);
     _____;
     printf("面积 s 是%f",s);
}
float area(float a,float b)
{    float s1;
     s1=a*b;
     _____;
}
```

(2) 以下函数 invert 的功能是,利用递归法实现对一个整数的逆序输出。

```
#include<stdio.h>
void invert(int n)
{    if(n<10)printf("%d",n);
     else
     {   printf("%d",n%10);
         _____;
     }
}
```

```
void main()
{   int x;
    scanf("%d",&x);
    invert(x);
}
```

(3) 从键盘上输入一个整数 n,输出 n 项对应的斐波那契数列。斐波那契数列是一整数数列,该数列第一项为 0,第二项为 1,从第三项开始,每个数等于前面两数之和。

```
#include<stdio.h>
int fib(int n);
void main()
{   int i,n=0;
    printf("请输入 n:");
    scanf("%d",&n);
    for(i=0;i<n;i++)
    printf("%5d",fib(i));
}
int fib(int n)
{   if(_____) return 0;
    else if(_____)return 1;
    else return _____;
}
```

(4) 将华氏温度转换为摄氏温度,转化公式为 C=(5/9)*(F-32),其中 C 为摄氏温度,F 为华氏温度。

```
#include<stdio.h>
void main()
{   float ctem,ftem;
    float _____;
    scanf("%f",&ftem);
    if(ftem<300)
        _____;
    else printf("超出了范围!\n");
    printf("摄氏温度是%f\n",ctem);
}
float trans(float tem)
{
    _____;
}
```

(5) 下列函数 fun 的功能是:用辗转相除法求正整数 m1 和 m2 的最大公约数,并返回该值。

```
#include<stdio.h>
int fun(int m1,int m2)
```

```
{   int t,a,b;
    if(m1 _____ m2)
    {   t=m1;
        _____;
        m2=t;
    }
    a=m1;   b=m2;
    while(_____)
    {   t=a%b;   a=b;   b=t; }
    return a;
}
void main()
{   int x,y;
    printf("请输入两个整数:");
    scanf("%d%d",&x,&y);
    printf("公约数为%d",_____);
}
```

（6）以下程序中函数 max 的功能是求两个数中的最大值，在主函数中输入两个数，然后调用函数 max 求得最大值并输出。

```
#include<stdio.h>
int max(int x,int y)
{   int z;
    z=_____;
    return z;
}
void main()
{   int a,b;
    scanf("%d%d",&a,&b);
    c=_____;
    printf("max is %d\n",c);
}
```

（7）以下函数的功能是：将输入的一个偶数写成两个素数之和的形式。例如，若输入数值 8，则输出 8=3+5。

```
#include<stdio.h>
#include<math.h>
void fun(int a)
{   int b,c,d;
    for(b=3;b<=a/2;b=_____)
    {   for(c=2;c<sqrt(b);c++)
            if(b%c==0)break;
        if(c>sqrt(b))d=_____;
        else break;
        for(c=2;c<=sqrt(d);c++)
```

```
            if(d%c==0)break;
        if(c>sqrt(d))printf("%d=%d+%d\n",a,b,d);
    }
}
void main()
{   int a;
    printf("\n intput a:\n");
    scanf("%d",&a);
    fun(a);
}
```

(8) 以下函数 fun 的功能是：统计一个数中位值为零的个数以及位值为 1 的个数。若输入 111001，则输出位值为零的个数为 2，位值为 1 的个数为 4。

```
#include<stdio.h>
void fun(long n)
{   int c1=0,c2=0,m;
    do{
        m=_____;
        if(m==0)c0++;
        if(m==1)c1++;
        n=_____;
    }while(n);
    printf("c0=%d,c1=%d\n",c0,c1);
}
void main()
{   long n;
    printf("\ninput n:\n");
    scanf("%ld",&n);
    printf("n=%ld\n",n);
    fun(n);
}
```

(9) 下面程序的功能是计算以下分段函数的值。

$$y = \begin{cases} 2.5-x & 0 \leqslant x < 2 \\ 2-1.5(x-3)^2 & 2 \leqslant x < 4 \\ \dfrac{x}{2}-1.5 & 4 \leqslant x < 6 \end{cases}$$

```
#include<stdio.h>
double y(_____)
{   if(x>=0&&x<2)   return (2.5-x);
    else if(x>=2&&x<4) return (2-1.5*(x-3)*(x-3));
    else if(x>=4&&x<6) return (x/2.0-1.5);
}
void main()
{   float x;
```

```
        printf("Please enter x:");
        scanf("%f",&x);
        if(_____)
            printf("f(x)=%f\n",y(x));
        else printf("x is out!\n");
}
```

(10) 以下程序的功能是应用下面的近似公式计算 e^x 的值。

$$e^x = 1 + x + \frac{x^2}{2!} + \frac{x^3}{3!} + \cdots \quad (前 20 项的和)$$

其中函数 fun1 用来计算每项分母的值,函数 fun2 用来计算每项分子的值。例如,当 x=3 时,$e^x \approx 20.0855$。

```
#include<stdio.h>
float fun1(int n)
{   if(n==1) return 1;
    else _____;
}
float fun2(int x,int n)
{   int i;
    float j=1;
    for(i=1;i<=n;i++)
        _____;
    return j;
}
void main()
{   float exp=1.0;
    int n,x;
    printf("请输入一个数 x:");
    scanf("%d",&x);
    exp=exp+x;
    for(n=2;n<=19;n++)
        exp+=_____;
    printf("e 的 x 次方的值为%8.4f",exp);
}
```

提示:函数 fun1 的功能是求分母 n!(n=2,3,4,…)的值,用的是递归的方法。函数 fun2 的功能是求分子 x^n 的值,通过循环累乘即可。利用公式求前 20 项的累加和在主函数中进行,前 2 项先累加到和 exp 中,从分母是 2!那一项开始到最后一项共 18 项,通过循环,每循环一次累加一项加数,每项加数又是由分子除以分母构成,而分子和分母可以通过调用函数得到。

2. 改错题

下列各个程序中"/*****************/"的下一行中有错误,请仔细阅读程序,并根据题意改正。

(1) 计算并输出下列级数的前 n 项的和 Sn,直到 Sn 大于 q 为止,q 的值通过形参传入。

$Sn=2/1+3/2+4/3+\cdots+(n+1)/n$，例如，当 q 的值为 50.0 时，则函数的值为 50.416687。

```c
#include<stdio.h>
/***************/
double fun(double q);
{   int n;
    double s=0;
    n=1;
    /***************/
    while(s>=q)
    {   /***************/
        s=s+(n+1)/n;
        n++;
    }
    printf("n=%d\n",n);
    return s;
}
void main()
{   double sum;
    sum=fun(50);
    printf("前50项的和为：%f",sum);
}
```

(2) 求出一个长整型数据的各位数字之积，积作为函数值返回。长整型数据由键盘输入。

```c
#include<stdio.h>
#include<math.h>
void main()
{   long n;
    /***************/
    long func(long num)
    printf("请输入一个数：\n");
    /***************/
    scanf("%d",&n);
    printf("各位数字的积是%ld",func(n));
}
long func(long num)
{   long k=1;
    num=abs(num);                    /* abs 为求绝对值的标准数学函数 */
    do{ /***************/
        k*=num/10;
        /***************/
        num%=10;
    }while(num);
    return(k);
```

}
```

提示：①求整型数各位数字的积必须先分离出各个数位，累乘起来。方法是：用该整数除以 10 取余数得到其个位数，用于求积；再用该整数整除 10 取商得到新的整数，用于继续分解各位数，直到商为 0。②func 函数的调用直接作为 printf 函数的输出项。

(3) 下列程序中函数 fac 的功能是：应用递归法求某数 a 的平方根。求平方根的迭代公式为 $x_n = \frac{1}{2}\left(x_{n-1} + \frac{a}{x_{n-1}}\right)$，要求直到前后两次求出的 x 差的绝对值小于 $10^{-5}$ 为止。

```
#include<stdio.h>
#include<math.h>
/************************/
fac(double a,double x0)
{   double x1,y;
    x1=(x0+a/x0)/2;
    /***********************/
    if(fabs(x1-x0)<1e-5)y=fac(a,x1);
    else y=x1;
    return y;
}
void main()
{   double x;
    printf("enter x:");
    scanf("%lf",&x);
    printf("%lf 的平方根是%lf\n",x,fac(x,1.0));
}
```

(4) 用二分法求方程 $2x^3 - 4x^2 + 3x - 6 = 0$ 在[m,n]内的一个根，m 和 n 从键盘上输入，要求绝对误差不超过 0.001。例如，若给 m 输入 -100，给 n 输入 90，则函数求得的一个根为 2.000。

```
#include<stdio.h>
#include<math.h>
double funx(double x)
{
    return (2*x*x*x-4*x*x+3*x-6);
}
double fun(double m,double n)
{   double r;
    r=(m+n)/2;
    /***************/
    while(fabs(n-m)<0.001)
    {   if(funx(r)*funx(n)<0) m=r;
        else n=r;
        r=(m+n)/2;
    }
```
```

```
        return r;
}
void main()
{   double m,n,root;
    printf("请输入 m 和 n:");
    /*************************/
    scanf("%f %f",&m,&n);
    root=fun(m,n);
    printf("root=%6.3f\n",root);
}
```

(5) 函数 fun 的功能是求 20 以内所有 5 的倍数之积。

```
#include<stdio.h>
void main()
{   int sum, fun(int m);
    sum=fun(5);
    printf("20 以内所有 5 的倍数之积为：%d\n",sum);
}
int fun(int m)
{   /********************/
    int s=0,i;
    for(i=1;i<20;i++)
        /********************/
        if(i%m==0)
            /*******************/
            s= * i;
    return s;
}
```

(6) 函数 fun 的功能是输入两实数，求它们的平方根之和，作为函数值返回。

```
#include<stdio.h>
#include<math.h>
/************************/
double fun(double a,double b);
{   double c;
    /**********************/
    c=sqr(a)+sqr(b);
    return c;
}
void main()
{   double a,b,y;
    printf("请输入两个实数：");
    /***********************/
    scanf("%lf%lf",a,b);
    y=fun(a,b);
```

```
    printf("这两个实数的平方根之和为%lf\n",y);
}
```

(7) 函数 decomp 的功能是,将一个正整数分解质因数。例如,输入 90,打印出 90＝2*3*3*5。

```
#include<stdio.h>
void decomp(int n)
{   int i;
    for(i=2;i<=n;i++)
    {   /******************/
        while(n==i)
        {   /*********************/
            if(n%i==1)printf("%d * ",i);
            /*********************/
            n=n%i;
        }
    }
}
void main()
{   int a;
    printf("请输入一个正整数: \n");
    scanf("%d",&a);
    printf("%d=",a);
    decomp(a);
}
```

(8) 有 5 个人坐在一起,问第 5 个人的岁数,他说比第 4 个人大两岁。问第 4 个人的岁数,他说比第 3 个人大两岁。问第 3 个人,又说比第 2 个人大两岁。问第 2 个人,说比第 1 个人大两岁。最后问第 1 个人,他说是 10 岁。请问第 5 个人多大?

```
#include<stdio.h>
int age(int n)
{   int c;
    /*************************/
    if(n=1) c=10;
    /*************************/
    else c=age(n)+2;
    return c;
}
void main()
{   /************************/
    printf("%d",age5);
}
```

(9) 计算并输出 k 以内最大的 10 个能被 13 或 17 整除的自然数之和。k 的值由主函数输入。

```
#include<stdio.h>
int fun(int k)
{   int m=0,mc=0,j;
    /***********************/
    while((k>=2)||(mc<10))
    {   /***********************/
        if(k%13=0||(k%17=0))
        {   m=m+k;    mc++; }
        /***********************/
        k++;
    }
    return m;
}
void main()
{   int k;
    printf("请输入一个自然数：");
    scanf("%d",&k);
    printf("%d\n",fun(k));
}
```

(10) 用公式：$\pi/4 = 1 - 1/3 + 1/5 - 1/7 + \cdots$ 求圆周率 π 的近似值，直到最后一项的绝对值小于或等于 10^{-4}。

```
#include<stdio.h>
#include<math.h>
void fun(int m)
{   int i;
    /***********************/
    int s=0,t=1,p=1;
    /***********************/
    while(fabs(t)<=1e-4)
    {   s=s+t;
        p=-p;
        i=i+2;
        t=p/i;
    }
    /***********************/
    printf("pi=%d\n",s*4);
}
void main()
{   fun(); }
```

3. 编程题

(1) 编写函数用来求表达式 $x^2 - 4x + 9$，x 作为参数传送给函数。

(2) 编写函数，若有参数 year 为闰年，则返回 1，否则返回 0。

(3) 编写函数 int fun(int x)，它的功能是判断整数 x 是否为同构数。若是同构数，函数返回 1；否则返回 0。所谓"同构数"是指这样的数：它出现在它的平方数的右边。例如，输入整数 5,5 的平方是 25,5 是 25 中右侧的数,所以 5 是同构数。x 的值由主函数从键盘读入,要求不大于 100。

(4) 编写函数,采用递归方法计算 x 的 n 次方。

(5) 编写函数,用递归算法将任一十进制整数转换成二进制数。

(6) 编写函数,能将十六进制数转换成十进制数。

(7) 编写一个判断素数的函数,在主函数输入一个整数,输出是否素数的信息。

(8) 编写函数,用递归方法求 n 阶勒让德多项式的值,递归公式为:

$$P_n(x) = \begin{cases} 1 & (n=0) \\ x & (n=1) \\ ((2n-1)xP_{n-1}(x) - (n-1)P_{n-2}(x))/n & (n>1) \end{cases}$$

(9) 输入一个正整数 x 和一个正整数 n,求下列算式的值。要求定义和调用 2 个函数,fact(n) 计算 n 的阶乘;pow(x,n) 计算 x 的 n 次幂(即 x^n),两个函数的返回值类型都是 double。

$$x - \frac{x^2}{2!} + \frac{x^3}{3!} - \frac{x^4}{4!} + \cdots + (-1)^{n-1} \frac{x^n}{n!}$$

(10) 给出某年某月某日,计算该日是该年的第几天。

(11) 编写函数 fun 求 s=a+aa+aaa+⋯+(此处 a 和 n 的值在 1~9 之间)。例如,a=3,n=5,则表达式为:s=3+33+333+3333+33333。在主函数中输入 a 和 n 的值,通过调用 fun 函数完成计算。

三、实验步骤

本实验要求掌握函数的应用,以一个具体实例说明函数的定义、调用方式,函数参数的传递方式,被调用函数将返回值带回主调函数的方法。

1. 题目:编写函数 fun(n),判断 n 是否为水仙花数,是返回 1,否返回 0。编写 main( ) 函数,输入一个数 num(在 100 到 999 之间),调用 fun(num) 函数,输出判断结果(水仙花数是指其各个数位的立方和等于该数本身,例如:$371 = 3^3 + 7^3 + 1^3$,则 371 为一水仙花数)。

2. 算法分析

根据题目要求,可以把一个三位数 n 的个位、十位、百位分解出来,假设为 i,j,k,若 $i^3 + j^3 + k^3$ 与 n 相等,则 n 为水仙花数。主函数从 100~1000 通过循环调用该函数,若返回值为 1,则将当前的数输出到屏幕上。

3. 根据分析,写出如下代码:

```
#include<stdio.h>
int fun(int n)                          /*函数形参为整型变量 n*/
{   int a,b,c;
    a=n%10;b=n/10%10;c=n/100;
    if((a*a*a+b*b*b+c*c*c)==n)return 1; /*满足条件返回 1*/
    else return 0;                      /*不满足条件返回 0*/
}
void main()
```

```
{   int num;
    scanf("%d",&num);
    while(num<100||num>=1000)
    {   printf("请重新输入,输入的数应为三位自然数");
        scanf("%d",&num);
    }
    if(fun(num)==1)                    /*调用函数 fun,实参为整型变量 num*/
        printf("%d是水仙花数\n",num);
    else printf("%d不是水仙花数\n",num);
}
```

4. 上机调试

将源程序代码录入到 Visual C++ 代码编辑窗口中,然后进行编译、连接,排查其中可能存在的错误,无误后运行程序。程序运行三次,分别输入不同的数查看运行结果,如图 9.1～图 9.3 所示。由输出结果可知,满足题意要求。

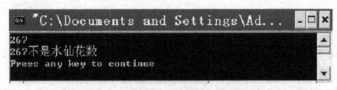

图 9.1　程序运行第一次输出结果

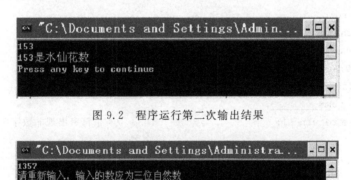

图 9.2　程序运行第二次输出结果

图 9.3　程序运行第三次输出结果

四、思考题

1. 对于 C 语言程序中的函数,函数的定义能嵌套吗? 函数调用能嵌套吗?
2. 什么是递归调用? 函数递归调用过程如何? 请写出程序填空中递归调用的全过程。
3. 函数一定有参数吗?
4. 函数在调用时值是如何传递的?
5. 在 C 语言函数中,可以没有 return 语句吗? 可以有多条 return 语句吗?

实验十 函 数（二）

一、实验目的与要求

1. 熟练掌握用数组元素和数组名作函数参数时，形参和实参的对应关系。
2. 熟练掌握参数的"值传递"以及"地址传递"的方式。
3. 掌握变量作为函数参数、数组作为函数参数的混合运用。
4. 掌握函数的定义及在实际编程中的应用。

二、实验内容

1. 填空题

请根据题意在下面各程序中划线处填写适当的语句或表达式，使之能够运行并获得正确的结果。

（1）编写函数，把数字字符串转换成整数。例如输入字符串12345，应输出整数12345。

```
#include<stdio.h>
int inter(char str[])
{   int i,num=0;
    for(i=0;str[i]!='\0';i++)
    {   if(str[i]<='9'&&str[i]>='0')
            num=_____;
        else return -1;                    /*字符串出现非数字字符输入出错*/
    }
    _____;
}
void main()
{   char str[80];
    scanf("%s",str);
    if(inter(str)==-1)printf("输入的字符串中有非数字字符,出错\n");
    else printf("%d\n",inter(str));
}
```

（2）计算输入的字符串中字母个数。要求以数组元素作为函数实参。

```
#include "stdio.h"
int  isalp(char c)
{   if(_____)return(1);
    else return(0);
}
void main()
```

```
{   int i,num=0;
    char str[255];
    printf("Input   a   string: ");
    gets(str);
    for(i=0;_____;i++)
        if(isalp(str[i]))                    /*调用函数,数组元素作为实参*/
            num++;
    puts(str);
    printf("字母的个数 num=%d\n",num);
}
```

(3) 编写一个函数 delchar(s,c),将字符串 s 中出现的所有 c 字符删除。编写 main 函数,并在其中调用 delchar(s,c)函数。

```
#include<stdio.h>
void delchar(char a[],char c)
{   int i,j;
    for(i=0; _____ ; i++)
        if(a[i]==c)
        {   for(j=i+1;a[j]!='\0';j++)
                a[j-1]=a[j];
            a[j-1]=_____; i--;
        }
}
void main()
{   char s[80],ch;
    printf("请输入字符串:\n");
    gets(s);
    printf("输入要删除的字符:\n");
    scanf("%c",&ch);
    delchar(_____);
}
```

(4) 以下程序中,函数 sum 的功能是求出 M 行 N 列二维数组每列元素中的最小值,并计算它们的和值。

```
#include<stdio.h>
int sum(int a[2][4])
{   int i,j,k,s=0;
    for(i=0;i<4;i++)
    {   k=0;
        for(j=1;j<2;j++)
            if(a[k][i]>a[j][i])k=j;
        s+=_____;
    }
    return s;
}
```

```
void main()
{   int x[2][4]={3,2,5,10,7,1,8,9},s;
    s=sum(_____);
    printf("%d\n",s);
}
```

(5) 设计一个函数 fc，其功能为统计数组中偶数的个数，编写 main 函数，用数组名 num 作为函数传递的参数调用 fc 函数，实现对数组 num 的统计，并输出统计结果。

```
#include<stdio.h>
int fc(int a[],int n)
{   int i,c=0;
    for(i=0;i<=n;i++)
        _____;
    return c;
}
void main()
{   int i,num[10];
    for(i=0;i<10;i++)
        scanf("%d",&num[i]);
    printf("偶数个数为：%d\n",_____);
    printf("奇数个数为：%d\n",_____);
}
```

(6) 在一个一维数组中存放 10 个正整数，找出其中所有的素数（用数组元素作为函数的实际参数）。

```
#include<stdio.h>
int sushu(int x)
{   int i,k=1;
    if(x==1) k=0;
    for(i=2;i<=x/2;i++)
        _____;
    return k;
}
void main()
{   int a[10],i;
    printf("请输入数组:\n");
    for(i=0;i<10;i++)
        _____;
    printf("数组中的素数个数为：\n");
    for(i=0;i<10;i++)
        if(_____) printf("%5d",a[i]);
    printf("\n");
}
```

(7) 设计一个函数,使输入的一个字符串按反序存放,在主函数中输入和输出字符串。

```
#include<stdio.h>
#include<string.h>
void main()
{   _____;
    char str[100];
    printf("请输入字符串:\n");
    gets(str);
    _____;
    printf("输出反序后的字符串:\n");
    _____;
}
int inverse(char st[])
{   char t;
    int i,j;
    for(i=0,j=strlen(st);i<strlen(st)/2;i++,j--)
    {   t=st[i];
        st[i]=st[j-1];
        st[j-1]=t;
    }
}
```

(8) 设计一个函数,输入一个 4 位数字,要求输出这 4 个数字字符,但每个数字字符间空一个空格。如输入 2012,应输出"2 0 1 2"。

```
#include<stdio.h>
#include<string.h>
void main()
{   void insert(char str[]);
    printf("请输入一个四位数字:\n");
    gets(str);
    _____;
    printf("输出插入空格后的数字:\n");
    puts(str);
}
void insert(char str[])
{   int i;
    for(i=strlen(str);i>0;i--)
    {   str[2*i]=str[i];
        str[2*i-1]=' ';
    }
}
```

2. 改错题

下列各个程序中"/**************/"的下一行中有错误,请仔细阅读程序,并根据题意改正。

(1) 输入10个数，输出这10个数的平均值。要求用数组名作函数的实际参数。

```
#include<stdio.h>
float average(float array[10])
{   int i;
    float aver,sum=array[0];
    /***********************/
    for(i=0;i<10;i++)
        sum=sum+array[i];
    aver=sum/10.0;
    return(aver);
}
void main()
{   /***************************/
    int score[10],aver;
    int i;
    printf("输入10个数：\n");
    for(i=0;i<10;i++)
        /***************************/
        scanf("%f",score);
    printf("\n");
    /***************************/
    aver=average(score[10]);
    printf("这10个数的平均值是%5.2f",aver);
}
```

(2) 用起泡法对10个整数从小到大排序。

```
#include<stdio.h>
/***************************/
void sort(int x,int n)
{   int i,j,k,t;
    for(i=0;i<n-1;i++)
        /***************************/
        for(j=0;j<n-i;j++)
        /***************************/
        if(x[i]>x[i+1])
        {   t=x[j]; x[j]=x[j+1]; x[j+1]=t;   }
}
void main()
{   int i,n,a[100];
    printf("请输入数组的长度：\n");
    scanf("%d",&n);
    for(i=0;i<n;i++)
        scanf("%d",&a[i]);
```

```
              /***************************/
              sort(n,a);
              printf("输出排序后的数：\n");
              for(i=0;i<n;i++)
                  printf("%5d",a[i]);
              printf("\n");
         }
```

(3) 编写一个函数,该函数可以统计一个长度为 3 的字符串在另一个字符串中出现的次数。例如输入的字符串为：trstrsdefghtrszx34trsklmp,子字符串为 trs,则应输出 4。

```
         #include<stdio.h>
         #include<string.h>
         int fun(char str[],char sub[])
         {   /***************************/
             int i,n=0
             /***************************/
             for(i=0;i<=strlen(str);i++)
                 if((str[i]==sub[0])&&(str[i+1]==sub[1])&&(str[i+2]==sub[2]))
                     /***************************/
                     ++i;
             return n;
         }
         void main()
         {   char str[80],sub[4];
             int n;
             printf("请输入主字符串：\n");
             gets(str);
             printf("请输入子字符串：\n");
             gets(sub);
             n=fun(str,sub);
             printf("子字符串在主字符串中出现的次数为%d\n",n);
         }
```

(4) 先将在字符串 s 中的字符按正序存放到 t 串中,然后把 s 中的字符按逆序连接到 t 串的后面。例如,当 s 中的字符串为"ABCDE"时,则 t 中的字符串应为"ABCDEEDCBA"。

```
         #include<stdio.h>
         #include<string.h>
         #include<conio.h>
         void fun(char s[],char t[])
         {   int i,sl;
             sl=strlen(s);
             /***************************/
             for(i=0;i<sl;i++)
                 t[i]=s[i];
             for(i=0;i<sl;i++)
```

```
        /*****************************/
        t[sl+i]=s[sl-i];
        /*****************************/
        t[sl]='\0';
}
void main()
{   char s[100],t[100];
    printf("请输入字符串: \n");
    scanf("%s",s);
    fun(s,t);
    printf("结果是: %s\n",t);
}
```

(5) 输入一批整数存储在数组中,以 0 作为结束标志,计算数组元素中值为正数的平均值(不包括 0)。例如,数组中元素的值依次为 37,－45,24,4,－9,16,0,则程序的运行结果为 20.25。

```
#include<stdio.h>
double fun(int s[])
{   /************************/
    int sum=0.0;
    int c=0,i=0;
    /************************/
    while(s[i]=0)
    {   if(s[i]>0) sum+=s[i];
        c++;
        i++;
    }
    /**********************/
    sum\=c;
    /**********************/
    return c;
}
void main()
{   int x[1000];int i=0;
    do{
        scanf("%d",&x[i]);
    }while(x[i++]!=0);
    printf("%f\n",fun(x));
}
```

(6) 编写一个函数,输入一个十六进制数,输出相应的十进制数。

```
#include<stdio.h>
int htoi(char s[])
{   int i,n;
    n=0;
```

```
        for(i=0;s[i]!='\n';i++)
        {   if(s[i]>='0'&&s[i]<='9')
                /*************************/
                n=n*16+s[i];
            if(s[i]>='a'&&s[i]<='f')
                /*************************/
                n=n*16+s[i]-'a';
            if(s[i]>='A'&&s[i]<='F')
                n=n*16+s[i]-'A'+10;
        }
        return n;
}
void main()
{   int c,i,flag,flag1;
    char t[1000];
    i=0;    flag=0;
    printf("请输入一个十六进制数：\n");
    while((c=getchar())!='\0'&&i<1000)
    {   if(c>='0'&&c<='9'||c>='a'&&c<='f'||c>='A'&&c<='F')
        {   flag=1;
            /*************************/
            t[i]=c;
        }
        else if(flag)
        {   t[i]='\0';
            printf("相应的十进制数为%d\n",htoi(t));
        }
        i++;
    }
}
```

（7）在函数 count 中，由实参传来一个字符串首地址，统计此字符串中字母、数字、空格和其他字符的个数，在主函数中输入字符串并输出结果。

```
#include<stdio.h>
/*************************/
int let,dig,spa,oth;
void count(char str[])
{   int i;
    for(i=0;str[i]!='\0';i++)
    if(str[i]>='a'&&str[i]<='z'||str[i]>='A'&&str[i]<='Z')
        let++;
    /*************************/
    else if(str[i]>='0'||str[i]<='9')dig++;
    /*************************/
    else if(str[i]=32)spa++;
```

```
        else oth++;
}
void main()
{   char text[80];
    printf("请输入字符串：\n");
    gets(text);
    count(text);
    printf("字母:%d,数字:%d,空格:%d,其他字符:%d\n",let,dig,spa,oth);
}
```

(8) 编写一个函数，使给定的一个二维数组(3×3)转置，即行列互换。

```
#include<stdio.h>
/************************/
void convert(int array[3])
{   int i,j,t;
    for(i=0;i<3;i++)
        for(j=i+1;j<3;j++)
        {   t=array[i][j];
            array[i][j]=array[j][i];
            array[j][i]=t;
        }
}
void main()
{   int i,j,array[3][3];
    printf("请输一个二维整型数组(3×3)：\n");
    for(i=0;i<3;i++)
        for(j=0;j<3;j++)
            /***************************/
            scanf("%d",array[i][j]);
    printf("原来数组为：\n");
    for(i=0;i<3;i++)
    {   /**********************/
        for(j=0;j<i;j++)
            printf("%5d",array[i][j]);
        printf("\n");
    }
    convert(array);
    printf("转置后的数组为：\n");
    for(i=0;i<3;i++)
    {   for(j=0;j<3;j++)
            printf("%5d",array[i][j]);
        printf("\n");
    }
}
```

(9) 计算个人与各科平均成绩及全班平均成绩，并在屏幕上显示出来。要求用二维数组名作为函数参数。定义一个(M+1)*(N+1)的二维数组,并进行初始化,留下最后一列 score[i][N]存放个人平均成绩,最后一行 score[M][i]存放学科平均成绩,最后一个元素 score[M][N]存放全班总平均。

```
#define M  5                                    /*定义符号常量 人数为5*/
#define N  4                                    /*定义符号常量 课程为4*/
#include<stdio.h>
void main()
{   int i,j;
    void aver(float sco[M+1][N+1]);
    static float score[M+1][N+1]={{78,85,83,65},{88,91,89,93},{72,65,54,75},
    {86,88,75,60},{69,60,50,72}};
    aver(score);                                /*调用函数,2维数组名作为实参*/
    printf("学生编号   课程1    课程2    课程3    课程4    个人平均\n");
    for(i=0;i<M;i++)
    {   printf("学生%d\t",i+1);
        for(j=0;j<N+1;j++)
            printf("%6.1f\t",score[i][j]);
        printf("\n");
    }
    printf("\n 课程平均");
    for(j=0;j<N+1;j++)
        printf("%6.1f\t",score[i][j]);
    printf("\n");
    getchar();
}
void  aver(float sco[][N+1])                    /*定义函数,二维数组名作为形参*/
{   int i,j;
    /*****************************/
    for(i=0;i<M+1;i++)
    {   for(j=0;j<N;j++)
        {   sco[i][N] +=sco[i][j];              /*求第 i 个人的总成绩*/
            sco[M][j] +=sco[i][j];              /*求第 j 门课的总成绩*/
            sco[M][N] +=sco[i][j];              /*求全班总成绩*/
        }
        /*****************************/
        sco[i][N]\=N;                           /*求第 i 个人的平均成绩*/
    }
    for(j=0;j<N;j++)
        /*****************************/
        sco[M][j]\=M;                           /*求第 j 门课的平均成绩*/
    sco[M][N]=sco[M][N]/M/N;                    /*求全班总平均成绩*/
}
```

3. 编程题

(1) 设有一个 3×4 的矩阵,求出其中的最大元素。要求:将计算最大值的功能设计成函数,数组名作为参数。

(2) 计算 1 门课程中 10 名学生成绩的平均分,要求编写函数实现。分数由主函数输入。

(3) 编写一个函数,判断一个字符串是否是回文字符串,如是返回 1,否则返回 -1(回文是指这个字符串逆置后不变,如 aba 就是回文字符串)。

(4) 从键盘输入一行字符串,由若干个英文单词组成,单词之间用空格分开,设计一个函数,返回该字符串中最长单词的长度,同时输出该单词的位置。

(5) 编写几个函数。①输入 10 个学生的姓名和学号;②按学号由小到大排序,姓名顺序也随之调整。从主函数输入要查找的学号,输出该学生姓名。

(6) 编写函数 fun 判断一个四位数是否满足条件:千位上的数减百位上的数减十位上的数减个位上的数大于 0。在主函数中调用该函数找出 5000 到 8000 之间满足条件的数的个数,并以每行 8 个输出满足条件的数及其该类数的个数。

三、实验步骤

本实验要求掌握数组名做函数参数、地址传递的应用方法。以具体实例说明数组名做函数参数与数组元素(作用同变量做函数参数)做函数参数的不同之处。

1. 题目:在主函数中输入两个正整数,调用一个函数求其最大公约数和最小公倍数,再调用一个函数将结果输出。

2. 算法分析

求两个数的最大公约数和最小公倍数的函数要同时返回两个值,按函数返回值的规则,它一次只能返回一个值,但这里需要两个值,因此,可以把这两个值存放在一个数组 p 里面,如 p[0]存放最大公约数,p[1]存放最小公倍数,通过数组的地址返回来达到要求。求最大公约数可以使用辗转相除法,前面已有介绍。本题中使用的方法是从两个数中较小者开始,以 1 为步长向下搜索,找到的第一个能同时被两个数整除的数即是它们的最大公约数。求出最大公约数后,将原来的两个数相乘再除以最大公约数,即得最小公倍数。

3. 根据分析,写出如下代码:

```
#include<stdio.h>
void cal(long num1,long num2,long p1[])      /*其中一个形参为数组名 p1*/
{   long k;
    k=num2>num1?num1:num2;
    while(k>1)
    {   if(num1%k==0&&num2%k==0) break;
        k--;
    }
    p1[0]=k;
    p1[1]=num1*num2/k;
}
void output(long p[])                         /*形参为数组名 p*/
{   printf("最大公约数为:%ld\n",p[0]);
```

```
        printf("最小公倍数为:%ld\n",p[1]);
}
void main()
{   long num1,num2,p[2];
    printf("请输入两个整数：\n");
    scanf("%ld%ld",&num1,&num2);
    cal(num1,num2,p);                    /* 调用函数 cal,其中一个实参为数组名 p */
    output(p);                           /* 调用函数 output,实参为数组名 p */
}
```

4. 上机调试

将源程序代码录入到 Visual C++ 代码编辑窗口中，接着进行编译、连接，排查可能存在的语法错误并纠正，无误后运行程序。程序运行结果如图 10.1 所示。

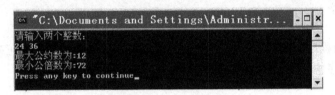

图 10.1　程序运行结果

四、思考题

1. 用数组名作函数参数与用普通类型变量作函数参数有什么不同？

2. 如何理解"用数组名作为函数实参时,形参数组中各元素的值如果发生变化会使实参数组元素的值同时发生变化"的含义。

3. 用数组元素做函数参数与普通变量做函数参数有什么不同？

实验十一 变量的存储类型与生存期

一、实验目的与要求

1. 掌握全局变量和局部变量、动态变量和静态变量的概念。
2. 熟练掌握全局变量和局部变量的使用方法。
3. 熟练掌握各种存储类型变量的定义方法、作用域及生存期。

二、实验内容

1. 阅读下列程序,分析程序的运行结果,并上机验证分析结果是否正确。

(1) 以下程序中使用了静态局部变量,注意静态变量的特点。

```
#include<stdio.h>
void fun()
{    static int a=0;                    /*定义静态局部变量a*/
     a+=2;
     printf("%d  ",a);
}
void main()
{    int cc;
     for(cc=1;cc<4;cc++)
         fun();
     printf("\n");
}
```

(2) 以下程序中使用了全局变量,请注意全局变量的特点。

```
#include<stdio.h>
int a,b;                                 /*定义全局变量a、b*/
void fun()
{    a=100;b=200;   }
void main()
{    int a=5,b=7;
     fun();
     printf("%d   %d\n",a,b);
}
```

(3) 以下程序中使用了全局变量,请注意全局变量与局部变量的区别。

```
#include<stdio.h>
int x=3;                                 /*定义全局变量x*/
```

```
void main()
{   int i;
    void incre();
    for(i=1;i<x;i++)  incre();
}
void incre()
{   static int x=1;
    x*=x+1;
    printf("%d  ",x);
}
```

(4) 以下程序中使用了静态局部变量,注意静态变量与动态变量的区别。若将静态局部变量改变为自动变量,分析程序运行结果有什么不同。

```
#include<stdio.h>
int f()
{   static int i=0;                    /*定义静态局部变量 i*/
    int s=1;
    s+=i; i++;
    return s;
}
void main()
{   int i,a=0;
    for(i=0;i<5;i++)
        a+=f();
    printf("%d\n",a);
}
```

(5) 以下程序中使用了相同名称的全局变量和局部变量,请注意它们的区别。

```
#include<stdio.h>
int d=1;
void fun(p)
{   int d=5;
    d+=p++;
    printf("%d  ",d);
}
void main()
{   int a=3;
    fun(a);
    d+=a++;
    printf("%d\n",d);
}
```

2. 改错题

下列各个程序中"/***************/"的下一行中有错误,请仔细阅读程序,并根据题意改正。

(1) 程序功能：利用全局变量计算长方体的体积及三个面的面积。

```c
#include<stdio.h>
int s1,s2,s3;
int vs(int a,int b,int c)
{   int v;
    v=a*b*c;  s1=a*b;  s2=b*c;  s3=a*c;
    /***************************/
    return  s1;
}
void main()
{   int v,l,w,h;
    printf("Input length,width and height: \n");
    /***************************/
    scanf("%d%d%d",l,w,h);
    v=vs(l,w,h);
    printf("v=%d   s1=%d   s2=%d   s3=%d\n",v,s1,s2,s3);
}
```

(2) 程序功能：输出 1~4 的阶乘。

```c
#include<stdio.h>
long  factorial(int );                /*函数声明*/
void main()
{   /***************************/
    int num=1;
    for(; num<=4; num++)
    printf("%d! =%ld\n", num, factorial(num));
}
long  factorial(int n)
{   /***************************/
    long fact=1;
    fact *=n;
    return fact;
}
```

(3) 以下程序在主函数 main 中输入一个数 5,通过函数调用使得该数的值加上 1 后,分别在主函数和被调用函数中输出。

```c
#include<stdio.h>
void add(int);
void main()
{   int num;
    /***************************/
    scanf("%d",num);
    add(num);
    printf("%d\n",num);                /*输出 5*/
```

```
}
void add(int num)
{   num++;
    printf("%d\n",num);                    /*输出 6*/
    /**************************/
    return num;
}
```

(4) 程序功能：求 1+2+…+100 的值。

```
#include<stdio.h>
void add();
int result;
void main()
{   register int i;
    /**************************/
    result=1;
    for(i=0;i<100;i++) add();
    printf("%d\n",result);
}
void add()
{   /**************************/
    static int num=1;
    num++;
    result+=num;
}
```

(5) 程序功能：关于外部变量的定义、声明与使用。

```
#include<stdio.h>
int vs(int xl,int xw)
{   extern int xh;                         /*外部变量 xh 的声明*/
    int v;
    v=xl*xw*xh;                            /*直接使用外部变量 xh 的值*/
    /**************************/
    return v;
}
void main()
{   extern int xw,xh;                      /*外部变量的声明*/
    int xl=5;                              /*内部变量的定义*/
    printf("xl=%d,xw=%d,xh=%d\nv=%d\n",xl,xw,xh,vs(xl,xw));
}
/****************************/
int xl=3,xw=4;xh=5;                        /*外部变量 xl、xw、xh 的定义*/
```

3. 编程题

(1) 编写一个函数 fun(a,n)，其中 a 是一维数组，n 是数组长度，要求通过全局变量

pave 和 nave 将数组中正数的平均值和负数的平均值传递给主函数后输出。数组各元素值在主函数中输入，调用 fun 函数时将数组的首地址和数组长度传递给形参。

（2）编写函数 seek(a,n)，从整型数组 a 中查找指定值 n，若找到，则将该值和它在数组中的位置传递给主函数后输出。a 和 n 均由主函数传递给 seek 函数。

（3）通过一个函数 fun，求 n 内所有能被 3 整除的自然数的个数和能被 7 整除的自然数的个数。

（4）有公式 $S=\dfrac{1}{1\times 2}+\dfrac{1}{2\times 3}+\cdots+\dfrac{1}{n\times(n+1)}$（例如当 n＝10 时，S＝0.909091），要求使用静态变量编写函数来求 S 的值。

（5）编写一个函数 fun，计算 n 门课程的平均分，查找最高分和最低分，将平均分、最高分和最低分传递给主函数。

三、实验步骤

本实验要求掌握全局变量的意义、全局变量与局部变量的区别，静态存储与动态存储的区别。举例说明全局变量的应用。

1. 题目：使用全局变量，编写程序实现通过一个函数 double calc(double x,double y)，可以得到两个实数的和、差、积、商。

2. 算法分析

本题只通过一个函数要求得到四个结果，而一个函数只能有一个返回值，所以可将两个数的和定为返回值求得，另外的差、积、商通过三个全局变量获得。三个全局变量在函数 calc 中得到计算结果，在主函数 main 中输出。

3. 根据分析，写出如下代码：

```
#include<stdio.h>
double a,b,c;                          /*定义全局变量*/
double calc(double x,double y)
{    a=x-y;
     b=x*y;
     c=x/y;
     return x+y;
}
void main()
{    double m,n;                       /*定义局部变量*/
     printf("请输入两个数：\n");
     scanf("%lf%lf",&m,&n);
     printf("%.2lf+%.2lf=%.2lf\n",m,n,calc(m,n));
     printf("%.2lf-%.2lf=%.2lf\n",m,n,a);
     printf("%.2lf*%.2lf=%.2lf\n",m,n,b);
     printf("%.2lf/%.2lf=%.2lf\n",m,n,c);
}
```

4. 上机调试

将源程序代码录入到 Visual C++ 代码编辑窗口中,然后进行编译和连接,排查程序中存在的语言错误并纠正,无误后运行程序。程序运行结果如图 11.1 所示。

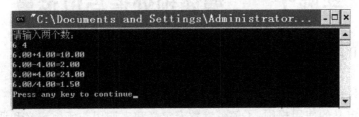

图 11.1　程序运行结果

四、思考题

1. 简述变量的分类方法。
2. 简述全局变量和局部变量的特点及作用域。
3. 静态变量和动态变量的区别是什么?

实验十二　编译预处理与位运算

一、实验目的与要求

1. 了解编译预处理命令。
2. 掌握宏定义的使用规则，了解带参数宏定义和宏调用与函数定义和调用的区别。
3. 了解文件包含并能熟练使用，了解常用函数的头文件。
4. 了解条件编译的用法，能使用条件编译编写程序，掌握条件编译和选择结构的异同。
5. 掌握位运算的运算规则，以及位运算对数据特定位的操作方法。

二、实验内容

1. 填空题

请根据题意在下面各程序中划线处填写适当的语句或表达式，使之能够运行并获得正确的结果。

(1) 以下程序中使用了带参数的宏定义，请注意参数的替换。

```
#include<stdio.h>
#define MIN(x,y)   (x)<(y)?x:y              /*带参数的宏定义*/
void main()
{   int a=5,b=2,c=3,d=3,t;
    t=MIN(a+b,c+d)*10;
    printf("%d\n",t);
}
```

(2) 以下程序中使用了无参数的宏定义。

```
#include<stdio.h>
#define PT 5.5
#define S(x)    PT*x*x
void main()
{   int a=1,b=2;
    printf("%4.1f\n",S(a+b));
}
```

(3) 以下程序中使用了带参数的宏定义。

```
#include<stdio.h>
#define MA(x)   x*(x-1)
void main()
{   int a=1,b=2;
```

```
        printf("%d\n",MA(1+a+b));
}
```

(4) 下面的宏定义是通过带参的宏定义来求圆的面积。

```
#include<stdio.h>
_____ PI   3.1415926
#define AREA(r) _____
void main()
{   float r;
    printf("请输入圆的半径：\n");
    scanf("%f",&r);
    printf("圆的面积为%f\n", _____);
}
```

(5) 以下程序功能为对给定的数据左移 m 位和右移 m 位(10>m>0)。

```
#include<stdio.h>
void move(unsigned num,int n);
void main()
{   unsigned int a=65437;
    int m;
    printf("请输入一个小于 10 大于 0 的数：\n");
    scanf("%d",&m);
    move(a,m);
}
void move(unsigned num,int n)
{   unsigned x,y;
    x=_____;
    y=_____;
    printf("%x 左移%d 的结果为：%x\n",num,n,x);
    printf("%x 右移%d 的结果为：%x\n",num,n,y);
}
```

2. 改错题

下列各个程序中"/***************/"的下一行中有错误，请仔细阅读程序，并根据题意改正。

(1) 以下程序功能为求一个圆的面积。

```
#include<stdio.h>
/****************************/
#define  PI   4;
void  main()
{   float  r;
    printf("请输入圆的半径：\n");
    /****************************/
    scanf("%f",r);
```

```
      printf("%d\n",PI*r*r);
}
```

(2) 以下程序根据两个变量的值判断其大小。

```
#include<stdio.h>
#define MAX(a,b)   a>b
/*************************/
#define EQU(a,b)   a=b
#define MIN(a,b)   a<b
void main()
{   int a,b;
    printf("请输入两个变量的值：\n");
    /*************************/
    scanf("%d%d",a,b);
    if(MAX(a,b))   printf("MAX\n");
    if(EQU(a,b))   printf("EQU\n");
    if(MIN(a,b))   printf("MIN\n");
}
```

(3) 输入一个整数,判断它能否被 3 整除,要求用带参的宏定义实现。

```
#include<stdio.h>
/*************************/
#define DIVBY3(m)   (m)/3==0
void main()
{   int m;
    printf("请输入一个整数：\n");
    scanf("%d",&m);
    /***********************/
    if(DIVBY3(m));printf("%d 能被 3 整除",m);
    else printf("%d 能不被 3 整除",m);
}
```

(4) 以下程序分别用函数和带参数的宏求出三个数中的最大值。

```
#include<stdio.h>
/******************************/
#define MAX(a,b)   a>b?a:b
int max3(int a,int b,int c)
{   int m;
    m=a>b?a:b;
    /****************************/
    m=c>m?m:c;
    return m;
}
void main()
{   int i,j,k;
```

```
    printf("请输入三个整数: \n");
    scanf("%d%d%d",&i,&j,&k);
    printf("1.调用宏定义得最大值为%d\n",MAX(i,MAX(j,k)));
    printf("2.调用函数得最大值为%d\n",max3(i,j,k));
}
```

(5) 定义一个带参数的宏，求两个整数相除的余数。

```
#include<stdio.h>
#define R(m,n)    (m)%(n)
void main()
{   /*********************/
    int m;n;
    printf("请输入两个整数: \n");
    /**************************/
    scanf("%d%d",&m,n);
    printf("%d 除以 %d 的余数为:%d\n",m,n,R(m,n));
}
```

(6) 以下程序为给出一个数的原码，求该数的补码。

```
#include<stdio.h>
void main()
{   unsigned short int a;
    unsigned short int get(unsigned short);
    printf("请输入一个八进制数: ");
    /************************/
    scanf("%d",&a);
    printf("结果是:%o\n",get(a));
}
unsigned short int get(unsigned short value)
{   unsigned short int z;
    /***********************/
    z=value|0100000;
    if(z==0100000)z=~value+1;
    else z=value;
    return z;
}
```

3. 编程题

(1) 从键盘输入圆的半径 r，输出圆的周长 s 和面积 area。圆的周长 $s=2\pi r$，圆的面积 $area=\pi r^2$，要求定义两个带参数的宏用来表示 s 和 area。

(2) 定义一个带参数的宏，其作用是使两个参数的值互换，并编写 C 源程序，输入两个数作为使用宏时的实参，输出实现交换后的两个值。

(3) 分别用自定义函数和带参数的宏定义实现求两个数的乘积的功能，并比较两种实现方法的区别。

(4) 给定年份 year，定义一个宏以判断该年份是否为闰年。

三、实验步骤

本实验要求掌握编译预处理命令以及位运算。举例说明带参数的宏定义和宏调用方法。

1. 题目：求三角形的面积公式为 area$=\sqrt{s(s-a)(s-b)(s-c)}$，其中：$s=\frac{1}{2}(a+b+c)$，a、b、c 为三角形的三边。定义两个带参数的宏，一个用来求 s，另一个用来求 area。

2. 算法分析

本题要求用两个带参的宏来实现，调用带参数的宏时须指明要替换的字符串，替换字符串的参数个数和宏定义时的参数个数相同。根据题目要求，可以将 $s=\frac{1}{2}(a+b+c)$ 定义为宏：♯define S(a,b,c) (a+b+c)/2，将面积 area$=\sqrt{s(s-a)(s-b)(s-c)}$ 定义为宏：♯define AREA(a,b,c) sqrt(S(a,b,c)*(S(a,b,c)-a)*(S(a,b,c)-b)*(S(a,b,c)-c))。

3. 根据分析，写出如下代码：

```
#include<stdio.h>
#include<math.h>
#define S(a,b,c)    (a+b+c)/2                            /*带参数的宏定义*/
#define AREA(a,b,c) sqrt(S(a,b,c)*(S(a,b,c)-a)*(S(a,b,c)-b)*(S(a,b,c)-c))
void main()
{   float a,b,c;
    printf("请输入三角形的三条边：");
    scanf("%f%f%f",&a,&b,&c);
    if(a+b>c&&a+c>b&&b+c>a)
        printf("三角形的面积为:%6.2f\n",AREA(a,b,c));    /*宏调用*/
    else printf("不能构成一个三角形\n");
}
```

4. 上机调试

将源程序代码录入到 Visual C++ 代码编辑窗口中，然后进行编译和连接，查找程序中可能存在的语法错误并纠正，无误后运行程序。程序运行结果如图 12.1 和图 12.2 所示。

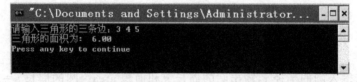

图 12.1　程序运行第一次输出结果

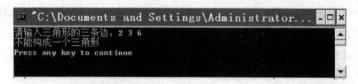

图 12.2　程序运行第二次输出结果

四、思考题

1. 在文件包含命令的使用中,"#include"后面的文件名用双引号和尖括号有什么区别?
2. 条件编译与选择结构有何不同?
3. 宏替换在何时进行?占用运行时间吗?

实验十三　综合实验(二)

一、实验目的与要求

1. 掌握一维数组和二维数组的定义方法、数组元素的引用方法以及程序设计的相关算法。
2. 掌握字符型数组和字符串函数在实际编程中的应用。
3. 熟练掌握函数的定义方法。掌握变量作为函数参数、数组作为函数参数的混合运用。
4. 掌握函数在实际编程中的应用。
5. 熟练掌握全局变量、局部变量的使用。

二、实验内容

1. 填空题

请根据题意在下面各程序中划线处填写适当的语句或表达式,使之能够运行并获得正确的结果。

(1) 有三行英文字符串,每行的长度不超过 10 个。以下程序中 fun 函数的功能是:将这三个字符串,按顺序合并成一个新的字符串。例如,若这三行中的字符串分别是:

AAA
BBBB
CCCCC

则合并后的字符串内容应该是 AAABBBBCCCCC。

```
#include<stdio.h>
#define M  3
#define N  10
void fun(char a[M][N],char b[])
{   int i,j,m=0;
    for(i=0;i<3;i++)
        for(j=0;a[i][j]!='\0';j++)
        {   b[m]=a[0][i];
            _____;
        }
        _____;
}
void main()
{   int i;
```

```
    char a[50];
    char w[M][N]={"AAA","BBBB","CCCCC"};
        printf("原三个字符串为：\n");
    for(i=0;i<M;i++)
        _____;
    printf("\n");
    fun(w,a);
    printf("合并后的字符串为：\n");
    printf("%s",a);
    printf("\n\n");
}
```

（2）用牛顿迭代法求方程的根。方程为 $ax^3+bx^2+cx+d=0$，系数 a、b、c、d 由主函数输入。求解 x 在 1 附近的一个实根。求出根后，由主函数输出。牛顿迭代法的公式是 $x_1=x_0-f(x_0)/f'(x_0)$，要求 x_1-x_0 差的绝对值小于 10^{-5} 为止。

```
#include<stdio.h>
#include<math.h>
float fun(float a,float b,float c,float d)
{   float x1=1,x0,f,f1;
    do
    {   x0=x1;
        f=((a*x0+b)*x0+c)*x0+d;
        f1=(3*a*x0+2*b)*x0+c;
        _____;
    }while(_____);
    return x0;
}
void main()
{   float a,b,c,d;
    printf("输入方程的系数 a,b,c,d：\n");
    scanf("%f%f%f%f",&a,&b,&c,&d);
    printf("方程是:%5.2fx^3+%5.2fx^2+%5.2fx+%5.2f=0\n",a,b,c,d);
    printf("\nx=%10.7\n",_____);
}
```

（3）以下程序中函数 fun 的功能是：将形参 a 所指数组中前半部分元素的值和后半部分元素的值对换。形参 n 中存放数组中数据的个数，若 n 为奇数，则中间的元素不动。

```
#include<stdio.h>
#define N 9
void fun(int a[],int n)
{   int i,t,p;
    p=(n%2==0)?n/2:_____;
    for(i=0;i<n/2;i++)
    {   t=a[i];
```

```
            a[i]=a[p+_____];
            _____=t;
        }
    }
    void main()
    {   int b[N]={1,2,3,4,5,6,7,8,9},i;
        printf("原数组为：");
            for(i=0;i<N;i++)
        printf("%4d",b[i]);
        printf("\n");
        _____;
        printf("对换后的数组为：");
        for(i=0;i<N;i++)
            printf("%4d",b[i]);
        printf("\n");
    }
```

(4) 程序功能：将十进制整数 n 转换为十六进制数，并将转换结果以字符串的形式输出。例如，输入十进制数 78，将输出十六进制数 4e。

```
#include<stdio.h>
#include<string.h>
char trans(int x)
{   if(x<10)return '0'+x;
    else return _____;
}
int DtoH(int n,char str[])
{   int i=0;
    while(n!=0)
    {   str[i]=trans(n%16);
        _____;
        i++;
    }
    return i-1;
}
void main()
{   int m;
    char str[100];
    printf("请输入一个十进制整数：\n");
    scanf("%d",&m);
    DtoH(m,str);
    printf("十进制数%d转换十六进制数为:%s\n",m,str);
}
```

(5) 以下程序中，函数 sumColum 的功能是求出 M 行 N 列二维数组每列元素中的最小值，并计算它们的和值。

```
#include<stdio.h>
#define M 2
#define N 4
int sum;
void sumColum(int a[M][N])
{   int i,j,k,s=0;
    for(i=0;i<N;i++)
    {   k=0;
        for(j=1;j<M;j++)
            if(a[k][i]>a[j][i]) k=j;
        s+=_____;
    }
    _____=s;
}
void main()
{   int x[M][N]={4,3,6,2,5,2,8,4},s;
    sumColum(x);
    printf("每列最小值的和为%d\n",sum);
}
```

2．改错题

下列各个程序中"/****************/"的下一行中有错误，请仔细阅读程序，并根据题意改正。

(1) 程序功能：利用顺序查找法从数组 a 的 10 个元素中对关键字 m 进行查找。

```
#include<stdio.h>
int a[10];
void main()
{   int m,i,no;
    int search(int a[],int);
    printf("请输入一个数组：");
    for(i=0;i<10;i++)
        scanf("%d",&a[i]);
    printf("请输入一个数 m:");
    scanf("%d",&m);
    /*************************/
    no=search(a[10],m);
    if(no!=-1) printf("找到,位置是%d\n",no+1);
    else printf("没找到\n");
}
int search(int a[10],int m)
{   int i;
    /*************************/
    for(i=0;i<9;i++);
        /*************************/
        if(a[i]=m) return i;
```

```
        return -1;
    }
```

(2) 以下程序中函数 fun 的功能是：从整数 1～55 之间，选出能被 3 整除且有一位数是 5 的数，并把这些数放在 b 所指的数组中，这些数的个数作为函数值返回。规定函数中 a1 放个位数，a2 放十位数。

```
#include<stdio.h>
/**************************/
int fun(int b[]);
{   int k,a1,a2,i=0;
    for(k=1;k<=55;k++)
    {   /************************/
        a2=k%10;
        a1=k-a2*10;
        if((k%3==0&&a2==5)||(k%3==0&&a1==5))
        {   b[i]=k;
            i++;
        }
    }
    /************************/
    return k;
}
void main()
{   int a[100],k,m;
    m=fun(a);
    printf("结果是：\n");
    for(k=0;k<m;k++)
        printf("%4d",a[k]);
    printf("\n");
}
```

(3) 在主函数中输入一个字符串，调用函数将其中所有的"is"替换成"be"，最后在主函数中输出结果。若子串"is"一个也没有找到，输出相应的提示信息。

```
#include<stdio.h>
int convert(char b[],char c[])
{   int i,count;
    while(b[i]!='\0')
    {   /************************/
        if((b[i]='i')&&(b[i+1]='s'))
        {   c[i++]='b';
            c[i++]='e';
            count++;
        }
        else { c[i]=b[i];  i++; }
    }
```

```
        c[i]='\0';
    /*************************/
    return i;
}
void main()
{   char a[50],b[50];
    int i,m;
    for(i=0;(a[i]=getchar())!='\n';i++);
    a[i]='\0';
    m=convert(a,b);
    /****************************/
    if(m==0) printf("%s\n",b);
    else printf("没找到\n");
}
```

(4) 主函数中输入一个 3×4 的整型矩阵,调用一个函数对其进行转置并将结果存放在另一个二维数组中。最后在主函数中输出结果。

```
#include<stdio.h>
void convert(int a[][4],int b[][3])
{   int i,j;
    for(i=0;i<3;i++)
        for(j=0;j<4;j++)
            b[j][i]=a[i][j];
}
void main()
{   int a[3][4],b[4][3];
    int i,j;
    for(i=0;i<3;i++)
        for(j=0;j<4;j++)
            /************************/
            scanf("%d",a[i][j]);
    convert(a,b);
    for(i=0;i<4;i++)
    {   for(j=0;j<3;j++)
            /***********************/
            printf("%d  ",a[i][j]);
        printf("\n");
    }
}
```

(5) 以下程序求:$S=2^1\times1!+2^2\times2!+\cdots+2^n\times n!(n<10)$的值,不使用数学函数而采用如下方法:先编写两个函数分别求解 2^n 和 n!,再编写求解 S 的函数,求解过程中调用前两个函数,最后在主函数中输入 n,调用求解 S 的函数完成任务。

```
#include<stdio.h>
long fun0(int);
```

```
long fun1(int);
long fun2(int);
void main()
{   int n;
    long result;
    printf("请输入计算的项数：\n");
    /************************/
    scanf("%d",n);
    result=fun0(n);
    printf("结果是:%ld",result);
}
    /************************/
long fun0(int n);
{   long result=0;
    if(n>1)
    {   result+=fun0(n-1);
        result+=fun1(n) * fun2(n);
    }
    else result+=2;
    return result;
}
long fun1(int n)
{   /************************/
    long result=0;
    int i;
    for(i=0;i<n;i++)
        result*=2;
    return result;
}
long fun2(int n)
{   long result;
    if(n>1)
    {   result=fun2(n-1);
        result*=n;
    }
    else result=1;
    return result;
}
```

3. 编程题

（1）编写函数 fun，它的功能是：求出 1～1000 内能被 7 或 11 整除，但不能同时被 7 和 11 整除的所有整数，并将它们放在 a 所指的数组中，通过 n 返回这些数的个数，然后在主函数中输出。

（2）编写函数 void fun(int m,int k,int ××[])，该函数的功能是：将大于整数 m 且紧靠 m 的 k 个素数存入×× 所指的数组中，在主函数中输入输出相应的数据。例如，若输入

11,5,则输出 13,17,19,23,29。

（3）有公式 $P=\dfrac{m!}{n!(m-n)!}$，m 与 n 为两个正整数且要求 m>n。编写两个函数,一个函数用来求一个数的阶乘,另一个用来求 P 的值,在主函数中输入 m 和 n 的值,并输出 P 的值。

（4）编写函数 int fun(int lim,int aa[MAX])，其功能是求出小于或等于 lim 的所有素数并放在 aa 数组中,该函数返回值为所求出的素数的个数。

（5）编写一个函数,用来删除字符串中所有的空格。例如,输入字符串"abc def ghi",输出"abcdefghi"。

三、实验步骤

1. 题目：在主函数中输入一个字符串,由若干英文单词组成,单词之间用空格分开。调用一个函数输出此字符串中最长的包含字母 a 的单词,若没有包含字母 a 的单词则输出相应的提示。

2. 算法分析

用于查找的函数可以使用一个循环,在输入的字符串中从头到尾逐一检索出每一个单词,同时检查其是否含 a 并求出单词的长度,检索出第一个含 a 单词时,将它存入一个辅助字符数组中,另用一个辅助变量记录它的长度,此后每次检索到含 a 的单词,即将它的长度与辅助数组中记录的单词长度比较,若发现当前检索到的单词更长,立即更新记录。如此操作直到输入的字符串结束,此时记录中的单词就是包含 a 的最长单词。

3. 根据分析,写出如下代码：

```
#include<stdio.h>
#include<string.h>
void find(char str[])
{   char word[50][50],word1[50][50];
    int i=0,j=0,k=0,m=0,n=0,leng=0;
    while(str[k]!='\n')
    {   if(str[k]==32)
        {   word[i][j]='\0';
            i++;
            j=0;
            k++;
        }
        else
        {   word[i][j]=str[k];
            k++;
            j++;
        }
    }
    for(k=0;k<=i;k++)
        if(strchr(word[k],'a'))
        {   j=0;
```

```
            while(word[k][j]!='\0')
            {   word1[m][j]=word[k][j];
                j++;
            }
            word1[m][j]='\0';
            m++;
        }
        if(m!=0)
        {   leng=strlen(word1[0]);
            for(k=0;k<m-1;k++)
                if(leng<strlen(word1[k+1]))
                {   n=k+1;
                    leng=strlen(word1[k+1]);
                }
            printf("%s\n",word1[n]);
        }
        else printf("没找到包含字符a的单词\n");
}
void main()
{   char str[50],c;
    int i=0;
    printf("请输入一个字符串：\n");
    while((c=getchar())!='\n')
    {   str[i]=c;
        i++;
    }
    str[i]='\0';
    find(str);
}
```

4. 上机调试

将源程序代码录入到 Visual C++ 代码编辑窗口中，然后进行编译和连接，排查程序可能存在的语法错误并纠正，无误后运行程序。程序运行结果如图 13.1 和图 13.2 所示。

图 13.1　程序运行第一次输出结果

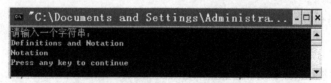

图 13.2　程序运行第二次输出结果

实验十四 指针与字符串

一、实验目的与要求

1. 了解指针的概念,掌握指针变量的定义、赋值及其应用。
2. 正确使用数组的指针和指向数组的指针变量。
3. 正确使用字符串的指针和指向字符串的指针变量。
4. 掌握指针变量作为函数参数的方法。

二、实验内容

1. 填空题

请根据题意在下面各程序中划线处填写适当的语句或表达式,使之能够运行并获得正确的结果。

(1) 利用指针变量找出 3 个数中的最小值并输出。指针变量应先赋值再使用。

```
#include<stdio.h>
void main()
{   int a,b,c,min;
    int *pa,*pb,*pc;                        /*定义指针变量*/
    printf("请输入 3 个整数:");
    scanf("%d%d%d",&a,&b,&c);
    _____
    _____
    pc=&c;
    if(_____)min=*pa;
    else min=*pb;
    if(*pc<min)min=*pc;
    printf("最小值为:%d\n",min);
}
```

(2) 利用指针变量输入 10 个数保存在数组中,并输出。如何使指针变量指向一个数组?如何使指针变量指向下或上一个元素?注意数组越界问题。

```
#include<stdio.h>
void main()
{   int i,a[10];
    _____
    p=a;
    printf("请输入 10 个整数:");
```

```
    for(i=0; i<10; _____)
        scanf("%d",p);
_____
    for(i=0;i<10;i++,p++)
        printf("%d,",*p);
    printf("\n");
}
```

(3) 利用指针变量将键盘输入的 3 个整数按照由小到大的顺序排序并输出。注意如何利用指针变量进行两个数据的大小比较。

```
#include<stdio.h>
void main()
{   int a,b,c,temp;
    int *pa,*pb,*pc;
    printf("请输入 3 个整数：");
    scanf("%d%d%d",&a,&b,&c);
    pa=&a; pb=&b; pc=&c;
    if(_____)
    {  temp=*pa; *pa=*pb; *pb=temp;  }
    if(_____)
    {  temp=*pa; *pa=*pc; *pc=temp;  }
    if(_____)
    {  temp=*pb; *pb=*pc; *pc=temp;  }
    printf("排序后:%d %d %d\n", _____);
}
```

(4) 利用函数调用完成：将键盘输入的 3 个整数按照由小到大的顺序排序并输出。在被调用函数 sort 中利用指针变量进行排序，sort 函数无返回值。注意函数调用时实参与形参的内容与形式，如何在无返回值的情况下实现实参数据排序。

```
#include<stdio.h>
void main()
{   int a,b,c;
_____
    printf("请输入 3 个整数：");
    scanf("%d%d%d",&a,&b,&c);
_____
    printf("调用函数排序后:%d %d %d\n",a,b,c);
}
void sort(_____)
{   int temp;
    if(*pa>*pb)
    {  temp=*pa; *pa=*pb; *pb=temp;  }
    if(*pa>*pc)
    {  temp=*pa; *pa=*pc; *pc=temp;  }
    if(*pb>*pc)
```

```
{ temp=*pb; *pb=*pc; *pc=temp; }
}
```

(5) 输入 10 个数,将其中的最小值与第一个数交换,将最大值与最后一个数交换。需要首先找到其中的最大值与最小值,并记录其位置,然后再进行交换。

```
#include<stdio.h>
void main()
{   int i,a[10],min,kmin,max,kmax,temp;
    int *p;
    p=a;
    printf("输入10个数:");
    for(i=0; i<10; i++,p++)
        scanf("%d",p);
    max=min=a[0];
    p=a;
    for(i=0;i<10;i++,p++)
    {   if(_____){ max=*p; kmax=i; }
        if(_____){ min=*p; kmin=i; }
    }
    temp=a[9]; a[9]=a[kmax]; a[kmax]=temp;     /*数据交换*/
    temp=a[0]; a[0]=a[kmin]; a[kmin]=temp;
    for(i=0,p=a;i<10; _____)
        printf("%d,",*p);
    printf("\n");
}
```

(6) 将一个 3 行 3 列的矩阵转置,即使 3×3 的二维数组行列互换。注意指针变量指向二维数组中某个元素时的格式。

```
#include<stdio.h>
void main()
{   int i,j,a[3][3],temp;
    int *p;
    p=_____;
    printf("请输入数据:\n");
    for(i=0; i<3; i++)
        for(j=0;j<3;j++)
            scanf("%d",&a[i][j]);
    for(i=0;i<3;i++)
        for(j=i;j<3;j++)
        {   temp=_____;
            _____=*(p+3*j+i);
            *(p+3*j+i)=temp;
        }
    printf("转置后:\n");
    for(i=0;i<3;i++)
```

```
        {   for(j=0;j<3;j++)
                printf("%d ",a[i][j]);
            printf("\n");
        }
}
```

(7) 两个字符串比较大小。从键盘输入两个字符串,若两个字符串不相等,则输出两个字符串中第一个不同字符的 ASCII 码值之差。

```
#include<string.h>
#include<stdio.h>
void main()
{   char str1[80],str2[80];
    int i=0;
    char *ps1=str1,*ps2=str2;
    printf("Input string 1:");
    gets(ps1);
    printf("Input string 2:");
    gets(ps2);
    while((ps1[i]==ps2[i])&&(ps1[i])!=_____) i++;
    printf("%d\n",_____);
}
```

(8) 从键盘输入 10 个整数,保存在数组中。查找数组中的最大值及其下标并输出。

```
#include<stdio.h>
void main()
{   int a[10],*p,*s,i;
    for(i=0;i<10;i++)
        scanf("%d",_____);
    for(p=a,s=a;_____<10;p++)
        if(*p>*s)s=_____;
    printf("max=%d,index=%d\n",_____,_____);
}
```

(9) 下列程序用来计算 N×N 阶方阵主、次对角线之和。

```
#include<stdio.h>
#define N 3
void sum(int x[][N],int *s1,int *s2)
{   int i,j;
    for(i=0;i<N;i++) *s1=*s1+_____;
    for(i=0;i<N;i++)
        for(j=N-1;j>=0;j--)
            if((i+j)==N-1) *s2=*s2+_____;
}
void main()
{   int a[N][N]={1,3,5,7,9,8,6,4,2};
```

```
    int s1=0,s2=0,i,j;
    sum(a,_____,_____);
    printf("主对角线之和:%d,次对角线之和:%d\n",s1,s2);
}
```

2. 改错题

下列各个程序中"/***************/"的下一行中有错误,请仔细阅读程序,根据题意改正程序中的错误,直至调试正确。

(1) 输入一个整数和一个实数,然后输出它们。

```
#include<stdio.h>
void main()
{   /*************************/
    int a,*pa;
    /*************************/
    float b,*pb;
    scanf("%d%f",pa,pb);
    printf("%d,%f\n",*pb,*pa);
}
```

(2) 利用指针变量为数组 a 输入 10 个整数,然后输出它们。

```
#include<stdio.h>
void main()
{   /*************************/
    int a[10],p,k;
    p=a;
    for(k=0;k<10;k++)
        scanf("%d",p++);
    /*************************/
    for(k=0;k<=10;k++)
        printf("%d ",*p++);
    printf("\n");
}
```

(3) 利用指针变量,将一个字符串复制到另一个区域中并输出。注意字符串拷贝函数的应用。

```
#include<stdio.h>
#include<string.h>
void main()
{   /*************************/
    char *str1,*str2="Hello world!";
    /*************************/
    strcmp(str1,str2);
    printf("%s\n",str1);
}
```

(4) 下面程序将数组 a 中的数据逆序存放并输出。数组名代表数组的首地址,也称为指针常量,其值不可改变。

```c
#include<stdio.h>
#define N 10
void main()
{   int a[N],i,j,temp;
    for(i=0;i<N;i++)
        scanf("%d",a+i);
    i=0;   j=N;
/**************************/
    while(i<j)
    {   temp=*(a+i);
/**************************/
        *(a+i)=*(a+j);
/**************************/
        *(a+j)=temp;
        i++;   j--;
    }
    for(i=0;i<N;i++)
        printf("%d ",a[i]);
    printf("\n");
}
```

(5) 下面程序用来判断输入的字符串是否为"回文串"(顺读和倒读都一样的字符串称为"回文串",例如 abcba 就为"回文串")。

```c
#include<stdio.h>
void main()
{   char  s[80],*t1,*t2;
    gets(s);
    t1=t2=s;
    while(*t2)t2++;
/**************************/
    while(t1<t2)
        if(*t1!=*t2)break;
        else   t1++;
/**************************/
    if(t1>t2)printf("Yes\n");
    else printf("No\n");
}
```

(6) 下面程序功能是,利用函数调用交换变量 a、b 的值。

```c
#include<stdio.h>
void swap(int *x,int *y)
{   int t;
```

```
    /************************/
    t=x;
    /************************/
    x=y;
    /************************/
    y=t;
}
void main()
{   int a=10,b=20;
    swap(&a,&b);
    printf("a=%d,b=%d\n",a,b);
}
```

(7) 将一个字符串添加到另一个字符串的末尾,然后输出新字符串。

```
#include<stdio.h>
void str_copy(char * s1,char * s2)
{   while(* s1)s1++;
    /************************/
    while(* s2)* s1=* s2;
    * s1=* s2;
}
void main()
{   /************************/
    char str1[]="abcd",str2[]="12345";
    str_copy(str1,str2);
    printf("复制后新字符串为:%s\n",str1);
}
```

(8) 下面程序将字符串中英文字母转换为按字母序列该字母的后一个字母,其他字母不变。例如,'a'转换为'b','b'转换为'c',……,而把'z'转换为'a'。

```
#include<stdio.h>
/************************/
void convert(char s)
{   do
    {   if(* s>='a'&&* s<='z') * s=(* s-'a'+1)%26+'a';
        if(* s>='A'&&* s<='Z') * s=(* s-'A'+1)%26+'A';
    /************************/
    }while(* s);
}
void main()
{   char str[]="abcdWZRTd26hy* 9nh";
    printf("转换前为:%s\n",str);
    convert(str);
    printf("转换后为:%s\n",str);
}
```

(9) 将从键盘输入的由数字字符组成的字符串转换为整数后输出。注意转换后的整数不要超出其取值范围。

```
#include<stdio.h>
int ctoi(char *s)
{   int sum=0;
    /**************************/
    while(s)
    {   sum*=10;
    /**************************/
        sum+=*s;
    }
    return sum;
}
void main()
{   char str[8];
    int n;
    printf("请输入数字字符串：");
    scanf("%s",str);
    /**************************/
    n=ctoi(&str);
    printf("转换为整数后：%d\n",n);
}
```

(10) 以下程序验证在 3~50 范围内，大于等于 3 的两个相邻素数的平方之间至少存在 4 个素数。例如，3^2 到 5^2 之间有素数 11、13、17、19、23。

```
#include<stdio.h>
#include<math.h>
int prime(int n)
{   int i;
    for(i=2;i<=sqrt(n);i++)
        if(n%i==0)return 0;
    return 1;
}
void main()
{   int i,j,k=0,m,n,count,a[50];
    int *p=a;
    for(i=3;i<50;i++)
        /**************************/
        if(prime(i))p[k]=i;
        /**************************/
    for(i=0;i<k;i++)
    {   m=p[i]*p[i];
        n=p[i+1]*p[i+1];
        count=0;
```

```
        for(j=m+1;j<n;j++)
            if(prime(j))count++;
        if(count>=4)
        printf("%d*%d--%d*%d:%d\n",a[i],a[i],a[i+1],a[i+1],count);
    }
}
```

3. 编程题(要求全部使用指针)

(1) 编写一个函数 itos(i,s),把一个整数转换为字符串存放到字符数组 s 中。要求在主函数中输入一个整数,调用函数 itos 实现整数到字符串的转换后,输出转换结果。

(2) 用梯形法编写程序求定积分 $f = \int_0^{10}(x^3 + x/2 + 1)dx$ 的值。

提示:用有限的矩形面积的和近似表示曲边形面积。设 b 为积分上限,a 为积分下限,n 为小矩形数量,每个小矩形的宽度为 $h = \frac{b-a}{n}$,则函数在(a,b)区间的定积分公式为 $s = \frac{h}{2}[f(a)+f(b)] + h\sum_{i=1}^{n-1}f(x_i)$。本题建议 n 为 1000,a 为 0,b 为 10。

(3) 统计一个字符串中包含的单词个数。例如字符串为:I am a student.,包含四个单词。

(4) 要求自定义函数比较两个字符串的大小。

提示:两个字符串比较大小,需要将两个字符串的对应位字符从左到右逐个进行比较,由字符 ASCII 码值的大小来定。假设串 1 为:"ABCDE",串 2 为:"12345","A"的 ASCII 码值为 65,"1"的 ASCII 码值为 49,则"A"大于"1",因此串 1 大于串 2。

(5) 对具有 m 个字符的字符串,从第 n 个字符开始的所有字符复制成为另一个字符串。要求字符串以及 n 都从键盘输入。例如,字符串为:"ncdjjufhinkjui",从第 5 个开始复制,新字符串为:"ufhinkjui"(不允许使用字符串处理函数 strcpy)。

(6) 有 n 位参赛者参加一个竞赛,每删除一个人之前生成一个随机数 num,然后按照顺时针方向数到 num,就删除之,直到剩下 s 位参赛者进入决赛。例如,有 20 人参加竞赛,最后只允许 10 位参赛者进入决赛。需要删除 10 人,首次生成的随机数设为 3,从第一个开始数,数到 3,删除对应的参赛者;再生成一个随机数,设为 5,从第四个开始数,数到 5,删除对应的参赛者;……,直至删掉 10 个为止。

(7) 字符串合并。编写一个函数 str_add(s,t,f),实现字符串 s 与字符串 t 的合并,f 作为标志,当 f 为 0 时,字符串时 s 连接到字符串 t 之后,当 f 为 1 时,字符串时 t 连接到字符串 s 之后。在主函数中输入两个字符串,确定 f 值,调用 str_add 函数实现字符串合并后,输出新串(不允许使用字符串处理函数 strcat)。

(8) 利用"筛法"求素数。方法是取一个从 2 开始的整数序列,通过不断划掉序列中非素数的整数(即合数)逐步确定序列中的素数。具体做法是:

① 令 n=2,2 是素数;
② 划掉序列中所有 2 的倍数;
③ 取 2 后面下一个未被划掉的数,即 3,再划掉后面所有 3 的倍数,依次类推,直至数列中所有整数都是素数。要求利用动态内存分配函数为预计的素数总数分配存储空间。

(9) 编写函数 void fun(*p,k,n),功能是移动 p 指向的一维数组中的元素。若该数组有 n 个元素,要求把下标从 k 到 n−1(k≤n−1)的数组元素平移到数组的前部。在主函数中输入数组各元素值和 k 值,调用 fun 函数实现移动后再输出数组。例如,一维数组的原始数据为:1,3,5,7,9,11,13,15,k 值为 3,则移动后数组元素应为:7,9,11,13,15,1,3,5。

(10) 求若干个整数的最小公倍数,正整数的个数由程序运行时输入的 n 值确定。要求利用动态内存分配函数为 n 个数据分配存储空间。

三、实验步骤

本实验要求掌握指针变量的应用。举例说明指针变量的定义、赋值、运算等。

1. 题目:从键盘输入一个英文字母,要求按照字母表的顺序输出 3 个相邻的字母,输入的字母在中间。例如,输入"A",则输出"ZAB";输入"Z",则输出"YZA"。

2. 算法分析

为了输出满足题目要求的连续 3 个英文字母,可以先定义字符数组,将 26 个字母按照字母表的顺序存储到数组中,定义指针变量并使其指向字符数组的起始位置,然后从键盘输入一个字符。关键的问题是找到该字符在数组中的前一个位置。

例如输入的字符为"D",利用表达式:p=p+('D'−'A')=p+3,得到的 p 值恰是该字符"D"在数组中的位置。本题目要求从前一个字符开始连续输出 3 个,则 p 值应减 1,即 p=p+('D'−'A'−1)。此外还要考虑,如果输入的字符处于数组的第一个或最后一个位置,需要做特殊处理。

当输入的字符为第一个字符时,那么,前一个字符应为数组的最后一个字符,即 *(str+25);当输入的字符为最后一个字符时,那么,下一个字符应为数组的第一个字符,此时应使指针指向数组头,即 p=str。

3. 根据分析,写出如下代码:

```
#include<stdio.h>
void main()
{    int i;
     Char ch, str[27]={"ABCDEFGHIJKLMNOPQRSTUVWXYZ"};
     char  *p=str;                              /*字符型指针变量定义并赋值*/
     ch=getchar();
     p=p+(ch-'A'-1);                            /*指针变量运算*/
     for(i=0;i<3;i++)
         if(p==str-1){printf("%c",*(str+25));p++;}
         else if(p!=str+26) printf("%c",*p++);   /*输出指针变量指向的数组元素值*/
         else { p=str; printf("%c",*p++);  }
     printf("\n");
}
```

4. 上机调试

将源程序代码录入到 Visual C++ 代码编辑窗口中,然后编译、连接,排查程序中可能存在的语法错误并纠正,无误后运行程序。程序运行三次,查看分别输入 3 个字符时运行结果是否满足题意要求。输入字符"A",输出"ZAB";输入字符"Z",输出"YZA";输入"S",输出"RST"结果均正确。程序运行结果如图 14.1～图 14.3 所示。

图 14.1　程序运行第一次输出结果

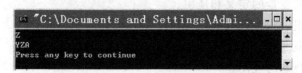

图 14.2　程序运行第二次输出结果

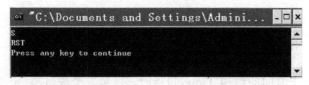

图 14.3　程序运行第三次输出结果

四、思考题

1. 整型的指针变量可以被赋予实型变量的地址吗？不同类型的指针变量可以互相赋值吗？
2. 一个指针变量可以被赋予非 0 的任意数值吗？
3. 指针变量可以进行的运算有哪些？与普通变量的运算有哪些区别？
4. 指针变量作为函数实参时,对应的形参可以是哪些类型？

实验十五　结构与联合

一、实验目的与要求

1. 掌握结构类型变量的定义和使用。
2. 掌握结构数组、结构指针的概念及其应用。
3. 掌握链表的概念，学会对链表进行基本操作。
4. 掌握联合的概念与应用。
5. 掌握枚举的概念及应用。

二、实验内容

1. 填空题

（1）定义一个关于时间的结构 Time，包括时、分、秒三个成员项，定义该结构类型变量，输入/输出一个具体时间。

```
#include<stdio.h>
_____                                       /*定义结构*/
{   int hour,minute,second;   };
void main()
{   struct Time t;                             /*定义结构变量*/
    printf("请输入某一时间,用时、分、秒描述:\n");
    scanf("%d%d%d", _____);
    printf("现在是%d时%d分%d秒.\n", _____);
}
```

（2）以下程序功能是输出结构类型数组 st 初值中的字母'M'，以及结构数组 st 所占据的内存单元个数。sizeof 可以检测数据类型、变量、结构、数组等所占据的内存单元数。

```
#include<stdio.h>
void main()
{   struct person                              /*定义结构*/
    {   char name[20];
        int age;
    };
    struct person st[4]={{"John",18},{"Paul",20},
                         {"Mary",19},{"Rose",20}};
    printf("%c\n", _____);
    printf("%d\n", _____);
}
```

(3) 定义一个结构变量,存储年、月、日等三个信息项。从键盘输入一个具体日期,计算该日是该年的第多少天。考虑闰年问题。

```
#include<stdio.h>
struct date                                              /*定义结构*/
{   int year, month, day;   };
void main()
{   int i,days;
    int day_tab[13]={0,31,28,31,30,31,30,31,31,30,31,30,31};
    struct date d;                                        /*定义结构类型变量*/
    printf("Enter year,month,day:");
    scanf("%d%d%d",_____);
    _____;
    for(i=1;i<_____;i++)
        days+=day_tab[i];
    if(d.year%4==0&&d.year%100!=0||d.year%400==0) _____;
    printf("%d/%d is the %dth day in %d\n",d.month,d.day,days,d.year);
}
```

(4) 在关于学生信息的结构数组中查找并输出最高分和最低分的同学姓名及成绩。

```
#include<stdio.h>
void main()
{   int max,min,i,j;
    struct
    {   char name[10];int score;}stu[5]={"Liping",99,"Lilin",89,
    "Wangfen",67,"Linhong",87,"Songhai",77};
    max=min=0;
    for(i=1;i<5;i++)
        if(stu[i].score>stu[max].score) _____;
        else if(stu[i].score<stu[min].score) _____;
    printf("Max is: %s,%d\n",_____);
    printf("Min is: %s,%d\n",_____);
}
```

(5) 输入 N 个整数,将输入的数及对应序号保存在结构中,并按照由小到大的顺序排列。当输入的 N 个数中有相同的数时,要求先输入的数排在前面。

```
#include<stdio.h>
#define N 10
struct input_num
{   int No;
    int num;
}array[N];
void main()
{   int i,j,n;
    for(i=0;i<N;i++)                                      /*输入数据并排序*/
```

```
    {   printf("Enter No.%d:",i);
        scanf("%d",&n);
        for(j=i-1;_____;j--)
            _____;
            _____;
        array[j+1].No=i;
    }
    printf("排序号 数据   输入序号\n");
    for(i=0;i<N;i++)                              /*输出排序后结果*/
        printf("%4d:%5d%5d\n",i,array[i].num,array[i].No);
}
```

(6) 一个班有 30 名学生,每个学生的数据包括学号、姓名和一门课的成绩。要求按学生的成绩由高到低排序,然后输出学生的信息以及平均成绩。

```
#include<stdio.h>
#define N 30
struct student
{   int num;  char name[20];   int score;   }stu[30];    /*定义结构数组*/
void main()
{   struct student *pt,*p[30];                    /*定义结构指针与指针数组*/
    int i,j,k,sum=0;
    for(i=0;i<N;i++)
    {   scanf("%d%s%d",&stu[i].num,stu[i].name,&stu[i].score);
        p[i]=&stu[i];
        sum=sum+_____;
    }
    for(i=0;i<N-1;i++)
    {   k=i;
        for(j=i;j<N;j++)
        if(_____) k=j;
        if(k!=i) { pt=p[i]; p[i]=p[k]; p[k]=pt;}
    }
    for(i=0;i<N;i++)
        printf("%d,%s,%d\n",_____);
    printf("Average=%d\n",_____);
}
```

(7) 下面程序的功能是,利用函数改变结构变量 a 的内容。a 包含两个成员项 x 和 c,x 原值为 10,改变为 20,c 原值为'x',改变为'y'。

```
#include<stdio.h>
struct stu
{   int x; char c;  };
void main()
{   void func(struct stu *);
    struct stu a={10,'x'},*p=&a;
```

```
         _____;
         printf("%d,%c\n",a.x, a.c);
}
void func(_____)
{    b->x=20;
     b->c='y';
}
```

(8) 下面程序是关于联合和联合变量的使用。联合中包括两个成员项,当为 x 成员项赋值后,可以由另一个成员项输出,并检测联合变量 a 占据内存单元的个数。

```
#include<stdio.h>
void main()
{   union                              /*定义联合*/
    {   int x;
        char c[2];
    }a;                                /*定义联合变量*/
    a.x=0x3456;
    printf("%c,%c\n",_____);        /*以字符形式输出 c[0]、c[1]的值*/
    printf("%x,%x\n",_____);        /*以十六进制数的形式输出 c[0]、c[1]的值*/
    printf("%d\n",_____);           /*检测联合变量占据内存单元数*/
}
```

(9) 在主函数中建立一个单向链表,head 指向单链表的第一个结点,调用函数 del 完成从链表中删除值为 num 的第一个结点。删除结点前、后分别输出链表。

```
#include<stdio.h>
#include<stdlib.h>
struct node
{   char ch;
    struct node * link;
}* head, * p, * q;
void main()
{   char c;
    struct node * del(struct node * h,char num);
    head=(struct node *)malloc(sizeof(struct node));
    p=q=head;
    c=getchar();
    p->ch=c;
    while((c=getchar())!='\n')         /*建立链表*/
    {    p=(struct node *)malloc(sizeof(struct node));
         p->ch=c;
         q->link=p;
         q=p;
    }
    q->link=NULL;
    p=head;
```

```
        while(p!=NULL)                          /*输出链表*/
        {   printf("%c ",p->ch);    p=p->link; }
        putchar('\n');
        c=getchar();                             /*输入需要删除的结点值*/
        del(head,c);                             /*调用函数 del 实施删除*/
        p=head;
        while(p!=NULL)                           /*输出删除结点后的链表*/
        {   printf("%c ",p->ch);    p=p->link; }
        putchar('\n');
}
struct node * del(struct node * h,char num)
{   struct node * p1,* p2;
    if(head==NULL)printf("list null!\n");        /*检测链表是否为空*/
    else
    {   p1=head;
        while(_____)                          /*在链表中查找要删除的结点*/
        {   p2=p1; p1=p1->link;    }
        if(num==p1->ch)
        {   if(p1==head) _____;               /*删除链表中值为 num 的第一个结点*/
            else _____;
            printf("delete:%c\n",num);
        }
        else printf("%c not been found!\n",num);
    }
    return head;
}
```

2. 改错题

下列各个程序中"/***************/"的下一行中有错误,请仔细阅读程序,根据题意改正程序中的错误,直至调试正确。

(1) 下面程序的功能是,准确输出联合类型变量 un 的两个成员项 a 和 x 的值。请注意,在联合类型的变量中某一时刻只有一个成员的值是有效的。要求说明错误原因。

```
#include<stdio.h>
void main()
{   union data                                   /*定义联合*/
    {   int a; float x;   };
    /**************************/
    data un;                                     /*定义联合变量*/
    un.a=5;
    /**************************/
    un.x=5.3;
    printf("%d\n",un.a);
    printf("%f\n",un.x);
}
```

(2) 定义枚举包含周日到周六的英文字头，字符型指针数组中存储周一到周六英文单词的首地址，从键盘输入 0～6 之间的整数，输出对应的英文单词。要求说明错误原因并纠正。

```
#include<stdio.h>
void main()
{   enum week{ sun,mon,tus,wed,thu,fri,sat }w;   /*定义枚举及枚举变量*/
    int n;
    /************************/
    char str[]={"Sunday","Monday","Tuesday","Wednesday",
                "Thursday","Friday","Saturday"};
    scanf("%d",&n);
    /************************/
    switch(n)
    {   case sun: w=sun; printf("%s\n",str[w]); break;
        case mon: w=mon; printf("%s\n",str[w]); break;
        case tus: w=tus; printf("%s\n",str[w]); break;
        case wed: w=wed; printf("%s\n",str[w]); break;
        case thu: w=thu; printf("%s\n",str[w]);break;
        case fri: w=fri; printf("%s\n",str[w]); break;
        case sat: w=sat; printf("%s\n",str[w]); break;
        default:printf("Input error.\n"); break;
    }
}
```

(3) 下面程序为联合变量的成员赋值，并依次输出各成员的值。要求说明错误原因。

```
#include<stdio.h>
void main()
{   union data                              /*定义联合及联合变量并初始化*/
    /************************/
    {   int n; char ch;float x; }comm={1,'a',3.5};
    printf("%d\n",comm.n);
    printf("%c\n",comm.ch);
    printf("%f\n",comm.x);
}
```

(4) 下面程序的功能是，为结构变量的成员赋值，并依次输出各成员的值，其中一个结构成员是联合类型变量。要求说明错误原因。

```
#include<stdio.h>
void main()
{   union a                                 /*定义联合*/
{   char c[6];float x; };
    struct st                               /*定义结构*/
    {   union a s;
        float y[5];
```

```
        double ave;
    }w;                                /*定义结构变量*/
    w.s.x=2.5;                         /*为结构变量成员赋值*/
    /**************************/
    w.s.c='*';
    /**************************/
    w.y=9.6;
    w.ave=1.3;
    printf("%f\n",w.s.x);              /*输出结构变量成员值*/
    /**************************/
    printf("%c\n",w.s.c);
    /**************************/
    printf("%f\n",w.y);
    printf("%f\n",w.ave);
}
```

(5) 根据输入的是星期几，计算出下一天是星期几并输出。考虑星期日与星期一的衔接问题。例如，输入星期日，那么，下一天应该是星期一。

```
#include<stdio.h>
#include<string.h>
enum day{ sun,mon,tue,wed,thu,fri,sat };    /*定义枚举类型*/
/**************************/
char week[]={"sun","mon","tue","wed","thu","fri","sat"};
void main()
{   char str[5];
    enum day d;
    printf("Input current date:");
    /**************************/
    scanf("%s",&str);
    if(strcmp(str,"sun")==0)d=sun;
    else if(strcmp(str,"mon")==0)d=mon;
    else if(strcmp(str,"tue")==0)d=tue;
    else if(strcmp(str,"wed")==0)d=wed;
    else if(strcmp(str,"thu")==0)d=thu;
    else if(strcmp(str,"fri")==0)d=fri;
    else if(strcmp(str,"sat")==0)d=sat;
    /**************************/
    printf("Tomorrow is %s\n",week[(d+1)]);   /*输出明天是星期几*/
}
```

(6) 下面程序能够正常执行，输出为："20，welcome"。请修改程序使它执行时输出"10，Hello!"。

```
#include<stdio.h>
struct wnum
{   int x;
```

```
    char * s;
}t;
/**************************/
void func(struct wnum t)
{   t.x=10;
    t.s="Hello!";
}
void main()
{   t.x=20;
    t.s="wellcome";
    /**************************/
    func(t);
    printf("%d,%s\n",t.x,t.s);
}
```

(7) 下列程序的功能是,建立一个带头结点的单向链表,新产生的结点总是插在表首,也就是说,第一个建立的结点在表尾。链表建立完毕,从表首开始输出。

```
#include<stdio.h>
#include<stdlib.h>
void main()
{   struct node
    {   char ch;
        struct node * link;
    } * h, * p;
    char c;
    h=NULL;
    while((c=getchar())!='\n')
    {   p=(struct node * )malloc(sizeof(struct node));
        p->ch=c;
        /**************************/
        h=p->link;
        /**************************/
        p=h;
    }
    p=h;
    while(p!=NULL)
    {   printf("%3c",p->ch);
        /**************************/
        p++;
    }
    putchar('\n');
}
```

(8) 设已经建立了一个链表,链表上结点的数据结构为:

struct node{float English, Math; struct node * next;};

求出该链表的结点个数、英语总成绩和数学总成绩,并在链首增加一个新结点,其成员 English 和 Math 分别存放这两门课程的平均成绩。若链表为空链时,链首不增加结点。函数 ave 的第一参数 h 指向链首,第二个参数 count 指向的空间存放求出的结点个数。

```c
#include<stdio.h>
#include<stdlib.h>
struct node                              /*定义结构*/
{   int English,Math;
    struct node * next;
} * h, * p, * q;                         /*定义结构指针*/
struct node * ave(struct node * h,int * count)
{   struct node * p1;
    int m=0,n=0;
    /*************************/
    if(h==NULL)return;                   /*检测链表是否为空*/
    p1=h;
    while(p1!=NULL)                      /*计算英语成绩总和、数学成绩总和*/
    {   m+=p1->English;
        n+=p1->Math;
        * count= * count+1;
    /*************************/
        p1->next=p1;
    }
    p1=(struct node * )malloc(sizeof(struct node));   /*建立一个新结点*/
    p1->English=m/( * count);            /*把英语平均分存入新结点*/
    p1->Math=n/( * count);
    /*************************/
    h=p1->next;
    /*************************/
    p1=h;
    return h;
}
void main()
{   int a,b,n=0;
    h= (struct node * )malloc(sizeof(struct node));
    p=q=h;
    scanf("%d%d",&a,&b);
    p->English=a;
    p->Math=b;
    while(1)                             /*建立链表*/
    {   p=(struct node * )malloc(sizeof(struct node));
        scanf("%d%d",&a,&b);
        if(a<=0&&b<=0)break;
        p->English=a;
```

```
            p->Math=b;
            q->next=p;
            q=p;
        }
        q->next=NULL;
        p=h;
        h=ave(h,&n);                          /* 调用 ave 函数 */
        printf("%d,%d\n",h->English,h->Math); /* 输出链表结点值 */
    }
```

3. 编程题

(1) 编写一个函数 t_f(),计算两个时刻之间的时间差,并将其值返回。在主函数中输入两个不同的时间,调用 t_f()求时间差后输出计算结果。要求时间以时、分、秒表示,两个时刻相差小于 24 小时。

(2) 已知某日是星期几,求 n 天后是星期几。

(3) 编写候选人得票的统计程序。设有 3 个候选人,每次输入一个得票的候选人名单,最后输出每个候选人的得票结果。参加投票的人数从键盘输入。

(4) 编写一个程序,使用结构数组存放下表中的数据,然后输出每人的姓名及应发工资数。扣税按照扣除标准 4000 元(包括三险一金)计算,在此基础上不超过 1500 元的部分速算扣除数为 0,超过 1500 元至 4500 元的部分速算扣除数为 105 元,超过 4500 元至 9000 元的部分速算扣除数为 555 元。速算扣除数即为扣税数,直接减去即可。

姓名	基本工资/元	岗位津贴/元	扣税/元	应发额/元
李红	3200	1200		
张华	5800	1800		
王青青	4680	1360		
李飞	7600	2300		
林丽	4560	1300		

(5) 口袋中有若干红、黄、蓝、白、黑 5 种颜色的球,每次从口袋中取出 3 个球,输出得到 3 种不同颜色球的所有可能取法(要求用枚举类型处理)。

(6) 某班有 30 人,期末考试科目有高等数学、英语、大学物理、程序设计,要求将所有学生按平均成绩排序,并输出四门课程均在 90 分以上的学生。定义结构表示学生的姓名和各科成绩,用结构数组保存全班学生的信息。

(7) 有若干个班,每个班人数不等,每个学生含姓名、成绩两个信息项。编写一个函数 aver_fun 统计各班平均成绩。在主函数中输入班级人数以及学生姓名和成绩,调用 aver_fun 计算平均分后,输出各班学生分数及平均分。

(8) 编程将链表逆转,即链首变成链尾,链尾变成链首。

(9) 编程求解 Josephus 问题。有 n(设 n=10)个小孩围成一圈,并给他们依次编上号 1~n,从第一个小孩开始报数,数到第 k 个孩子,该孩子出列,从下一个开始报数,依次重复下去,直至所有小孩出列。求所有小孩的出列顺序。要求采用结构数组存放小孩信息。

三、实验步骤

本实验要求掌握结构、联合、链表的应用。举例说明结构类型、结构变量的定义,引用结构变量中成员的方法等。

1. 题目:中国有句古话叫做"三天打鱼两天晒网"。某人从 2010 年 1 月 1 日起开始三天打鱼两天晒网,问这个人在以后的某一天是打鱼还是晒网。

2. 算法分析

根据题意,若要计算出某人在 2010 年 1 月 1 日之后的某一天是打鱼还是晒网,首先要计算从 2010 年 1 月 1 日到指定日有多少天,再按照三天打鱼两天晒网的规律,即 5 天一个周期,将计算出来的天数除以 5,根据余数可以得出结论。余 1、2、3 则打鱼,余 4 或能被 5 整除则晒网。

3. 根据分析,写出如下代码:

```c
#include<stdio.h>
void main()
{    int i=1,n,k,sum_day=0;
     struct date                              /*定义日期结构 date*/
     {   int year,month,day;   }d;            /*定义日期结构变量 d*/
     char m[]={0,31,28,31,30,31,30,31,31,30,31,30,31};
     printf("请输入指定日期(要求为 2010 年元旦后):\n");
     scanf("%d%d%d",&d.year,&d.month,&d.day);
     sum_day=sum_day+d.day;                   /*计算指定日期距 2010 元月 1 日的天数*/
     n=d.year-2010;
     k=d.year;
     if(n>=1)sum_day=sum_day+n*365;
     while(i<d.month){ sum_day=sum_day+m[i]; i++; }
     while(k>=2010)                           /*判断期间有无闰年*/
     {   if(k%4==0&&k%100!=0||k%400==0)sum_day++;
         k--;
     }
     n=sum_day%5;
     if(n==1||n==2||n==3)
         printf("%d年%d月%d日打鱼.\n",d.year,d.month,d.day);
     else printf("%d年%d月%d日晒网.\n",d.year,d.month,d.day);
}
```

4. 上机调试

将源程序代码录入到 Visual C++ 代码编辑窗口中,然后进行编译和连接,排查并纠正程序中可能存在的语法错误,无误后运行程序。运行结果如图 15.1 和图 15.2 所示。

图 15.1　程序第一次运行结果

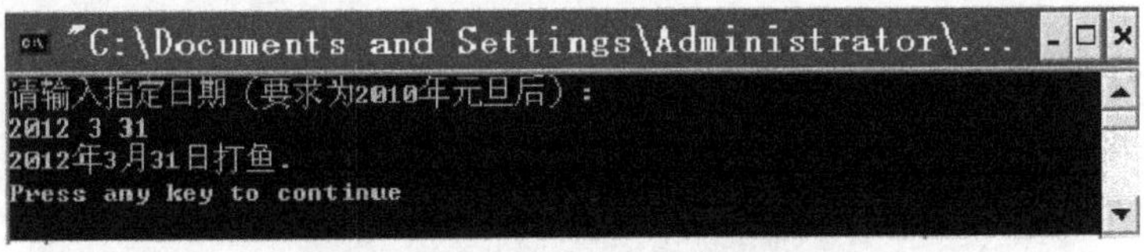

图 15.2　程序第二次运行结果

四、思考题

1. 结构与联合的主要区别是什么？
2. 结构与数组的主要区别是什么？
3. 枚举元素是常量还是变量？可以用任意类型数据为枚举变量赋值吗？为什么？
4. 在链表的使用中必须要用指针吗？为什么？

实验十六　排序与查找程序设计

一、实验目的与要求

1. 掌握排序算法及应用。
2. 掌握查找算法及应用。
3. 掌握插入算法及应用。
4. 掌握删除算法及应用。

二、实验内容

1. 填空题

请根据题意在下面各程序中划线处填写适当的语句或表达式,使之能够运行并获得正确的结果。

（1）删除数据。下面程序的功能是,调用函数 char_del,删除字符串中所有数字字符。函数中首先对逐个字符进行判断,若是数字字符,则删除。删除字符实际上是将该字符后面的字符依次前移。

```
#include<stdio.h>
#include<string.h>
#include<ctype.h>
void char_del(char * s)
{   int i=0;
    while(s[i]!='\0')
        if(isdigit(s[i])) _____;
        else _____;
}
void main()
{   char str[80];
    gets(str);
    _____;
    puts(str);
}
```

（2）数据排序。下面程序的功能是利用插入法将 n 个数从大到小进行排序。插入法排序的思想是：从一个空表开始,将待排序的数按照次序一个接一个地插入到已排好序的有序表中（空表视为有序表）,从而得到一个新的有序表。例如,有数据 3,−5,7,9,0,排序过程如下：

排序趟数	有序表	待排序数据
初始状态	空	3 −5 7 9 0
第一趟	3	−5 7 9 0
第二趟	3 −5	7 9 0
第三趟	7 3 −5	9 0
第四趟	9 7 3 −5	0
第五趟	9 7 3 0 −5	空

```
#include<stdio.h>
void sort(int a[],int b[],int n)
{   int i,j;
    b[0]=a[0];
    for(i=1,j=1;i<n;i++)                /*插入法排序*/
    {   j=_____;
        while(_____)
        {   b[j+1]=b[j];
            j--;
        }
        _____;
    }
}
void main()
{   int a[10],b[10],i;
    printf("Enter 10 numbers:\n");
    for(i=0;i<=9;i++)
        scanf("%d",&a[i]);
    sort(a,b,10);
    printf("Sorted:\n");
    for(i=0;i<10;i++)
        printf("%d ",b[i]);
    printf("\n");
}
```

(3) 查找数据。以下程序在数组 a 中查找与 x 值相同的元素所在位置。查找(或称为搜索)分顺序查找和二分查找,顺序查找适用于无序表,二分查找(也称为折半查找)适用于有序表。这里采用的方式是顺序查找。

```
#include<stdio.h>
void main()
{   int a[11],x,i;
    printf("Enter 10 integers:");
    for(i=1; i<=10;i++)
        scanf("%d",&a[i]);
```

```
    printf("Enter x:");
    scanf("%d",&x);
    a[0]=_____;
    i=10;
    while(x!=a[i]){ i--; if(i<=0){i++;_____;}}        /*顺序查找*/
    if(_____)printf("%5d 's position is:%4d\n",x,i);
    else printf("%d no found.\n",x);
}
```

　　(4) 删除重复数据。从数组开始依次选定一个元素,将该元素值与其后的所有元素相比较,如果相同就删除后者,之后的所有元素依次前移一个位置,并且将数组元素的个数减1。

```
#include<stdio.h>
#include<stdlib.h>
#include<time.h>
void main()
{   int a[10],n,i;
    void purge(int *pa, int *pn);
    srand(time(0));
    for(i=0; i<10;i++)
        a[i]=rand()%10;                /*用随机数函数产生10个数*/
    printf("原始数据为:\n");
    for(i=0;i<10;i++)
        printf("%d ",a[i]);
    n=10;
    purge(_____);
    printf("\n重复数据删除后:\n");
    for(i=0;i<n;i++)
        printf("%d ",a[i]);
    printf("\n");
}
void purge(int *pa,int *pn)              /*查找相同的数并删除*/
{   int i,j,k;
    for(i=0;i<*pn-1;i++)
    {   j=i+1;
        while(_____)
           if(pa[j]==pa[i])
           {   for(k=j+1;k<*pn;k++)
                   _____;              /*前移*/
               _____--;                /*数组元素个数减1*/
           }
           else j++;
    }
}
```

(5) 以下程序对一组点坐标(x,y)按升序进行排序。要求先按 x 的值排序,若 x 值相等再按 y 值排序。排序算法为选择法。

```
#include<stdio.h>
#define N 6
struct point{ int x; int y ;};
void Point_sort(struct point * s,int n)
{   struct point t;
    int i, j, k;
    for(i=0;i<n-1;i++)
    {   _____;
        for(j=_____;j<n;j++)
          if(s[k].x>s[j].x)k=j;
          else if (_____ && s[k].y>s[j].y)k=j;
        if(_____){ t=s[i]; s[i]=s[k]; s[k]=t;   }
    }
}
void main()
{   struct point a[N];
    int i=0;
    while(i<N)
    {   scanf("%d%d",&a[i].x,&a[i].y);
        i++;
    }
    Point_sort(a,N);
    for(i=0;i<N;i++)
      printf("%d,%d\t",a[i].x,a[i].y);
}
```

(6) 以下程序中 replace 函数的功能是,将字符串 s 中所有属于字符串 s1 中的字符都用 s2 中对应位置上的字符替换。例如,若 s 字符串为"ABACBA",s1 字符串为"AC",s2 字符串为"ac",则调用 replace 函数后,字符串 s 的内容将变为"aBacBa"。

```
#include<stdio.h>
#define MAX 20
void replace(char * s,char * s1,char * s2)
{   char * p;
    for(; * s;s++)
    {   p=s1;
        while(* p&& _____)p++;
        if(* p) * s=_____;
    }
}
void main()
{   char s[MAX]="ABBCFFBCSADQCBA",s1[MAX]="AC",s2[MAX]="ac";
    _____;
```

```
        printf("The string of s is:%s\n",s);
}
```

(7) 插入数据。程序功能是在有序数组中插入一个数,插入后原顺序保持不变。为此,需要将插入位置之后的数据后移一个位置。在数组中插入数据分两种情况,一是定点插入,即在指定位置插入一个数据;二是按顺序插入一个数据,这种方式要求原始数据就是按顺序排列的,插入新数据后原次序不变。一个数据插入后,数据个数增 1。

```
#include<stdio.h>
void main()
{   int i,j,n;
    int a[11]={-3,2,4,5,7,9,11,22,23,27};  /*初始值保留一个空位*/
    printf("Insert data:");
    scanf("%d",&n);
    for(i=0;i<10;i++)                       /*找插入位置*/
        if(_____) break;
    for(j=9;j>=i;j--)                       /*插入点之后的数从末尾开始顺序后移一位*/
        _____;
    _____;                               /*插入数据*/
    printf("New array:\n");
    for(i=0;i<=10;i++)
        printf("%d ",a[i]);
    printf("\n");
}
```

(8) 假设数组 a 中有 10 个元素,分别是 51、23、12、45、38、11、15、34、67、71。编写函数 Insert(),在数组中指定下标 k 处和插入 x 值,完成插入操作。在主函数中输入 k 和 x 值,并输出数组 a 插入数据后的各元素值。

```
#include<stdio.h>
void Insert(int a[],int k,int x,int n)
{
    ……
}
void main()
{   int a[11]={51,23,12,45,38,11,15,34,67,71};
    int i=0,k=0,x;
    printf("Please input position and number:\n");
    scanf("%d%d",&k,&x);
    Insert(a,k,x,10);
    for(i=0;i<11;i++)
        printf("%d ",a[i]);
    printf("\n");
}
```

2. 改错题

下列各个程序中"/***************/"的下一行中有错误,请仔细阅读程序,根据题意改正程序中的错误,直至调试正确。

(1) 下列程序在主函数中输入 10 个互不相同的整数存放于数组 a 中,调用 delete_max 函数删除其中的最大值,之后在主函数中输出数组 a 的其余 9 个数据。采用顺序查找的方式找出最大值。

```
#include<stdio.h>
/**************************/
int delete_max(int p)
{   int maxindex=0,maxvalue=0,i;
    for(i=1;i<10;i++)
    /**************************/
        if(p[i]<p[maxindex])maxindex=i;
    maxvalue=p[maxindex];
    for(i=maxindex;i<9;i++)
        p[i]=p[i+1];
    return maxvalue;
}
void main()
{   int a[10]={0},i=0,max=0;
    printf("Please enter 10 differnt integers:\n");
    for(i=0;i<10;i++)
    /**************************/
        scanf("%d",a);
    max=delete_max(a);
    printf("The max valu is:%d\n",max);
    printf("After deletint the maximum,the datum is:\n");
    for(i=0;i<9;i++)
        printf("%d ",a[i]);
    printf("\n");
}
```

(2) 下列程序的功能是对数组中存储的 n 个整数按照由小到大排序。排序算法是:第一趟将其中的最小值放在第一个位置,将最大值放在最后;第二趟将次小值放在第二个位置,次大值放在倒数第二个位置;……,依次类推。

```
#include<stdio.h>
#define N 10
void sort(int a[],int n)
{   int i,j,t,min,max;
    for(i=0;i<n/2;i++)
    /**************************/
    {   min=max=0;
        /**************************/
        for(j=i+1; j<n-1; j++)
            if(a[j]<a[min])min=j;
            else if(a[j]>a[max])max=j;
```

```
            if(min!=i)
            {   t=a[min]; a[min]=a[i]; a[i]=t;  }
            if(max!=n-i-1)
               if(max==i)
               {   t=a[min]; a[min]=a[n-i-1]; a[n-i-1]=t;  }
               else  {   t=a[max]; a[max]=a[n-i-1]; a[n-i-1]=t; }
        }
}
void main()
{    int a[N]={1,0,9,11,7,4,-3,22,-8,10},i;
     sort(a,N);
     /***************************/
     for(i=0;i<=N;i++)
        printf("%d ",a[i]);
     printf("\n");
}
```

(3) 函数 func 的功能是,删除字符串中的数字字符。在主函数中输入一个包含数字字符的字符串,调用 func 函数删除其中的数字字符后,输出新字符串。

```
#include<stdio.h>
#include<string.h>
void func(char * s);                              /*函数声明*/
void main()
{    char str[80];
     printf("Input string:");
     gets(str);                                    /*输入字符串*/
     func(str);                                    /*调用函数*/
     printf("After delete digitat char:%s\n",str); /*输出字符串*/
}
void func(char * s)
{    char * ps;
     /*************************/
     for(ps=s;ps!='\0';ps++)                       /*删除数字字符*/
     /*************************/
        while(*ps>='0'&&*ps<='9')strcpy(ps,ps-1);
}
```

(4) 函数 strnum(char * str,char * subst)的功能是,统计子串 subst 在字符串 str 中出现的次数。字符串 str 和子串 subst 在主函数中输入,调用函数 strnum 后输出子串在字符串中出现的次数(说明:函数 strncmp()与 strcmp()作用相当,只是前者多了一个参数 n。strncmp()用来判断子串与字符串中的指定个数的字符是否相同,它有三个参数,第一个是字符串,第二个是子串,第三个是每次判断的字符个数)。

```
#include<stdio.h>
#include<string.h>
```

```
int strnum(char * str,char * subst);         /*函数声明*/
void main()
{   char str[80],substr[20];
    printf("Input string:");
    gets(str);                               /*输入字符串*/
    printf("Input substring:");
    gets(substr);
    printf("\nn=%d\n",strnum(str,substr));
}
int strnum(char * str,char * subst)
{   int i=0,n=0,len1,len2;
    len1=strlen(str);
    len2=strlen(subst);
    /**************************/
    while(i<=len2)                           /*判断子串与字符串中的指定字符是否相同*/
    /**************************/
        if(strncmp(str+i,subst,len2)==0){ n++; i++; }
        else i++;
    return n;
}
```

(5) 函数 del() 的功能是, 从字符串中的指定位置删除 n 个字符。在主函数中输入一个字符串, 调用 del 函数后输出新串。

```
#include<stdio.h>
void del(char s[],int i,int n)
{   int j,k,length=0;
    while(s[length]!='\0')
        length++;
    --i;
    j=i;
    k=i+n;
    while(k<length)
    /**************************/
        s[k++]=s[j++];
    s[j]='\0';
}
void main()
{   int s,n;
    char str[80];
    printf("输入字符串:\n");
    gets(str);
    printf("输入指定位置以及要删除的字符个数:\n");
    scanf("%d%d",&s,&n);
    /**************************/
    del(str,n,s);
```

```
        printf("%s\n",str);
}
```

(6) 利用二分法在有序数组(例如由大到小排列)中查找指定数据。其操作过程是,先确定待查找元素的范围,确定中间元素的位置,然后将待查找的数据与中间元素值进行比较,若相等即查找到;若待查找数据大于中间元素值,则在前半部分查找;否则在后半部分查找。即将查找范围缩小一半,取这部分的中间位置,将该位置元素值与待查找数据比较,重复该过程,直至找到或查找范围缩小到 0,即未找到为止。

```
#include<stdio.h>
#define N 10
int main()
{   int a[N]={123,99,87,67,56,,45,23,11,-91,-100};    /*有序数组*/
    int m,found=0;
    int low=0,high=N-1,mid;
    printf("输入待查找的数:");
    scanf("%d",&m);
    /**************************/
    while(low>=high&&found==1)                          /*二分查找*/
    {   mid=(low+high)/2;
        /**************************/
        if(m>a[mid])low=mid+1;
        /**************************/
        else if(m<a[mid])high=mid-1;
        else found=1;
    }
    if(found==1)printf("找到%d!它位于下标为%d的位置。\n",m,mid);
    else printf("不存在该数.\n");
    return 0;
}
```

(7) 下面程序中函数 fun() 的功能是,将 s 所指向的字符串中除了下标为奇数、其 ASCII 码值也为奇数的字符,其余的字符都删除,串中剩余的字符所形成的新串存放在 t 数组中。例如,字符串为"mnhgt54rf",删除奇数位其值为非奇数的字符 n 和 r、偶数位上的字符 mht4f 后,新字符串为"g5"。

```
#include<stdio.h>
#include<string.h>
/**************************/
void fun(char s,char t[])
{   int k,n=0;
    for(k=0;k<strlen(s);k++)             /*删除偶数位及奇数位其ASCII值为非奇数的字符*/
    /**************************/
        if(k%2==1&&s[k]%2==1)t[++n]=s[k];
    t[n]='\0';
}
```

```
void main()
{   char s[80],t[80];
    printf("Please enter string:");
    gets(s);
    fun(s,t);
    printf("The new string is: %s\n",t);
}
```

(8) 输入 5 位用户的姓名(字符串)和电话号码(8 位数字),按照电话号码由大到小的顺序排列后输出。

```
#include<stdio.h>
#define N 5
struct user
{   char name[20];
    long num;
    /*************************/
} sp,temp;
void main()
{   int i,j,k;
    printf("Please enter string:");
    for(i=0;i<N;i++)
    /*************************/
        scanf("%s%ld",sp[i].name,sp[i].num);
    for(i=0;i<N-1;i++)
    {   k=i;
        for(j=i+1;j<N;j++)
        /*************************/
            if(sp[j].num>sp[k].num)k=j;
        if(k!=i){ temp=sp[k];sp[k]=sp[i];sp[i]=temp; }
    }
    printf("After sorted:\n");
    for(i=0;i<N;i++)
       printf("%20s,%ld\n",sp[i].name,sp[i].num);
}
```

3. 编程题

(1) 编写一个函数 insert(a,i,k),把整数 k 插入到数组 a 的第 i 个位置。在主函数中输入数组各元素值,调用 insert 函数插入指定数据后再输出(提示:先找到指定位置,将该位置以及其后的数据依次后移一个位置,空出该位置后,将 k 插入到该处)。

(2) 编写一个函数 getdata(int a[]),利用随机函数产生 50 个 10~99 之间的整数,存储在数组 a 中。编写函数 void sort(int a[],int b[]),将 a 数组中的偶数保存在 b 数组中,并将它们按照由小到大的顺序排序,排序方法为选择法。在主函数中分别定义长度为 50 的数组 a 和 b,用 a 作实参调用 getdata 函数,用 a 和 b 作实参调用 sort 函数,最后输出 b 数组排序后的结果。

(3) 利用插入法将上题中 getdata() 函数随机生成的 50 个整数按照由大到小顺序排列后输出。

(4) 输入若干姓名,按照英文字典顺序对这些姓名进行排序。

(5) 删除字符串中连续的相同的字符并输出。例如,字符串为:"programming",相同字符删除后应输出:"programing"。

(6) 假设字符串中只包含字母和 * 号。编写函数将字符串中所有的前导 * 号全部移到字符串的尾部。不得使用字符串处理函数。例如,原始字符串为:"***kldjfjurh**HFG",移动前导 * 号后字符串为:"kldjfjurh**HFG***"。

(7) 编写函数 move,将字符串中下标为奇数的字符移到下一个奇数的位置,右边被移出字符串的字符放到第一个奇数位置,下标为偶数的字符不移动。在主函数中输入一个字符串,调用 move 函数完成移动后,输出移动后的字符串。例如:

输入:string
移动后输出:sgrtni

(8) 编写程序,删除字符串中所有空格后输出新串。

三、实验步骤

本实验要求掌握常用排序方法、查找方法、插入与删除方法。举例说明 SHELL 法的排序过程。

1. 题目:利用 SHELL 法对一个数组的元素进行由小到大排序。SHELL 排序的思想是:设待排序的序列有 n 个元素,首先取一个整数 $k<n$ 作为间隔,将全部元素分为 k 个子序列,所有间隔为 k 的元素都为同一个子序列。在每一个子序列中分别采用插入算法排序,然后缩小 k,直到 $k=1$,并将所有对象放到同一个序列再排一次为止。由于开始时子序列对象较少,排序速度比较快,待到后期 k 值逐渐缩小,子序列中元素逐渐变多,但由于前面排序的基础,大多数数据已基本有序,再加上插入法的合理性,所以排序速度有一定的提高。

2. 算法分析

设有数据序列{9,0,3,5,7,2,1,8,6,4},这里有 10 个元素,取 $k=5$,相隔 5 个元素形成子序列,即 9 和 2、0 和 1、3 和 8、5 和 6、7 和 4 组成 5 个子序列,进行两两比较,其中 9 和 2、7 和 4 不满足升序要求,所以需要交换位置,其余不变,这是第一次排序;第二次取上次 k 值的二分之一,即 $k=2$,则 2、3、4、1 及 6 为子序列,0、5、9、8 及 7 为一个子序列进行排序,直至 $k=1$ 全部元素进行排序,即可得到最后的结果,如图 16.1 所示。

初始数据:	9	0	3	5	7	2	1	8	6	4
第一遍 k=5	2	0	3	5	4	9	1	8	6	7
第二遍 k=2	1	0	2	5	3	7	4	8	6	9
第三遍 k=1	0	1	2	3	4	5	6	7	8	9
排序结果	0	1	2	3	4	5	6	7	8	9

图 16.1 SHELL 法排序过程示意

3. 根据分析,写出如下代码:

```c
#include<stdio.h>
void shell(int count,int a[])                    /* SHELL法排序模块 */
{   int i,j,k=count,n,m;
    for(n=0;n<count;n++)                         /* 控制循环次数 */
    {   k=k/2;
        for(i=k;i<count;++i)                     /* SHELL法排序 */
        {   m=a[i];
            for(j=i-k;m<a[j]&&j>=0;j=j-k)
                a[j+k]=a[j];
            a[j+k]=m;
        }
        if(k==0)break;
        printf("k=%d:\n",k);                     /* 输出每次排序时的k值 */
        for(i=0;i<count;++i)                     /* 输出每次排序时的结果 */
            printf("%4d",a[i]);
        printf("\n");
    }
}
void main()
{   int i,n,a[20];
    printf("Input the n(n<=20):");
    scanf("%d",&n);                              /* 输入待排序数据个数 */
    for(i=0;i<n;i++)                             /* 输入待排序的数据 */
        scanf("%d",a+i);
    shell(n,a);                                  /* 调用SHELL函数 */
    printf("sorted:\n");
    for(i=0;i<n;++i)                             /* 输出排序后的结果 */
        printf("%4d",a[i]);
    printf("\n");
}
```

4. 上机调试

将源程序代码录入到 Visual C++ 代码编辑窗口中,然后进行编译、连接,排查并纠正程序中可能存在的语法错误,无误后运行程序。程序运行结果如图 16.2 所示。

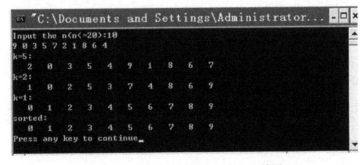

图 16.2 SHELL 法排序程序运行结果

四、思考题

1. 总结冒泡法排序、选择法排序、插入法排序的算法。
2. 顺序查找与折半查找算法的区别是什么？
3. 删除某个数据时是将该数据从内存中去掉吗？
4. 二分查找与顺序查找的区别是什么？

实验十七 文 件 操 作

一、实验目的与要求

1. 掌握文件、文件指针的概念。
2. 掌握利用函数打开、关闭、读、写文件的方法。
3. 掌握随机操作文件的基本方法。

二、实验内容

1. 填空题

请根据题意在下面各程序中划线处填写适当的语句或表达式,使之能够运行并获得正确的结果。

(1) 从键盘输入一个文件名,然后把从键盘输入的若干个字符写入到该文件中,用"♯"作为输入结束的标志。

```
#include<stdio.h>
#include<stdlib.h>
int main()
{   FILE * fp;                                  /*定义文件指针*/
    char ch,filename[10];
    printf("Input the name of file.\n");
    scanf("%s",filename);                       /*输入文件名*/
    if((fp=fopen(_____,_____))==NULL)     /*打开文件*/
    {   printf("Can't open file.\n");  exit(0);  }
    printf("Enter data.\n");
    while(_____)
        _____;                               /*写入文件*/
    fclose(fp);
    return 0;
}
```

(2) 从键盘输入一个字符串,将其中小写字母全部转换为大写字母,然后写入到一个文件 file1.dat 中保存,再从文件中读出并显示到屏幕上,检测该文件内容是否正确。

```
#include<stdio.h>
#include<stdlib.h>
void main()
{   FILE * fp;
    int i;
```

```
        char str[80];
        if((fp=fopen("file1.dat","w"))==NULL)        /*打开文件*/
        {  printf("can't open the file.\n");  exit(0);  }
        printf("Enter a string:");
        gets(str);                                    /*输入字符串*/
        for(i=0;str[i];i++)
        {  if(_____) str[i]-=32;
            _____;
        }
        _____;
        fp=fopen("file1","r");
        _____;
        printf("%s\n",str);
        fclose(fp);
}
```

(3) 已知 z=f(x,y)=(3.14*x－y)/(x＋y)，若 x、y 取值为区间[1,8]的整数,找出使 z 取最小值的 x1、y1,并将 x1、y1 以格式"%d,%d"写到文件 file2.dat 中。

```
#include<stdio.h>
void main()
{   FILE *p;
    float  min,_____;
    int x,x1,y,y1;
    p=fopen("file2.dat",_____);
    min=f(1,1);
    for(x=1;x<=8;x++)
        for(y=1;y<=8;y++)
            if(_____)
            { min=f(x,y); x1=x; y1=y; }
    fprintf(p,"%d,%d",x1,y1);
    fclose(p);
}
float f(float a,float b)
{   return (3.14*a-b)/(a+b);  }
```

(4) 将数组 a 中的每一行均除以该行上绝对值最大的元素,然后将 a 数组处理后的结果写到文件"file3.dat"中。

```
#include<stdio.h>
#include<math.h>
void main()
{   FILE *p;
    int i,j;
    float x,a[3][3]={{1.3,2.7,3.6},{2,3,-4.7},{2,5,1.27}};
    p=fopen("file3.dat","w");
    for(i=0;i<3;i++)
```

```
        {   x=0;
            for(j=0;j<3;j++)
                if(_____)x=fabs(a[i][j]);
            for(j=0;j<3;j++)
                _____;
        }
        for(i=0;i<3;i++)
        {   for(j=0;j<3;j++)fprintf(p,"%10.6f",a[i][j]);
            fprintf(p,"\n");
        }
        fclose(p);
    }
```

(5) 在 6～5000 之间找出所有的亲密数,并将所有亲密数用语句 fprintf(p,"%6d, %6d\n",a,b);写到文件"file4.dat"中(说明：若 a、b 为一对亲密数,则 a 的因子之和等于 b,b 的因子之和等于 a,并且 a 不等于 b。如 220 和 284 就是一对亲密数)。

```
#include<stdio.h>
void main()
{   FILE *p;
    int a,b,c,k;
    p=fopen("file4.dat","w");
    for(k=6;k<=5000;k++)
    {   b=0;c=0;
        for(a=1;a<k;a++)
            if(_____)b=b+a;
        for(a=1;a<b;a++)
            if(_____)c=c+a;
        if(_____)fprintf(p,"%6d,%6d\n",k,b);
    }
    fclose(p);
}
```

(6) 计算多项式 $a_0+a_1*\sin(x)+a_2*\sin(x*x)+a_3*\sin(x*x*x)+\cdots\cdots$ 的值,并将其值以格式"%.6f"写入到文件"file5.dat"中。

```
#include<stdio.h>
#include<math.h>
void main()
{   FILE *p;
    int i;
    double x=2.345,t=1.0,y=0;
    double a[10]={1.2,-1.4,-4.0,1.1,2.1,-1.1,3.0,5.3,6.5,-0.9};
    p=fopen("file5.dat","w");
    y=a[0];
    for(i=1;i<=9;i++)
    {   t=t*x;
```

```
        y=y+_____;
    }
    _____;
    fclose(p);
}
```

(7) 下面程序中函数 fun 的功能是,求出 1 到 m 之间能被 7 或 11 整除的整数并放到 a 数组中,通过 n 返回这些数的个数。

```
#include<stdio.h>
#define M 100
void fun(int m,int *a,int *n)
{   int k;
    for(k=1;k<=m;k++)
        if(k%7==0||k%11==0)a[(*n)++]=k;
}
void main()
{   FILE *p;
    int a[M],n=0,k;
    p=fopen("file6.dat","w");
    fun(_____);
    for(k=0;k<n;k++)
        if(k%10==0)printf("\n");
        else
        {   printf("%4d",a[k]);
            fprintf(p,"%d",a[k]);
            fputs(" ",p);
        }
    printf("\n");
    _____;
}
```

(8) 从键盘输入姓名,在文件 xm.dat 中查找,若文件中存在该姓名,则输出提示信息;若文件中没有该姓名,则将该姓名写入文件。要求:

① 若文件 xm.dat 存在,则保留原来的信息;若文件不存在,则建立该文件。

② 当输入的姓名为空时(即姓名字符串长度为 0),结束程序。

```
#include<stdio.h>
#include<string.h>
#include<stdlib.h>
void main()
{   FILE *fp;
    int flag;
    char name[30],data[30];
    if((fp=fopen("xm.dat",_____))==NULL)
    {   printf("Open file error\n"); exit(0);   }
```

```
    do
    {   printf("Enter name:");
        gets(name);
        if(strlen(name)==0)break;
        strcat(name,"\n");
        _____;
        flag=1;
        while(flag&&(fgets(data,30,fp)!=NULL))
            if(strcmp(data,name)==0)_____;
        if(flag)fputs(name,fp);
        else printf("Data found!\n");
    }while(ferror(fp)==0);
    fclose(fp);
}
```

2. 改错题

下列各个程序中"/***************/"的下一行中有错误，请仔细阅读程序，根据题意改正程序中的错误，直至调试正确。

(1) 下面程序是将用户从键盘输入的字符写入到文件"file7.dat"中，直至输入'!'为止，将'!'写入文件后关闭文件，程序结束。

```
#include<stdio.h>
void main()
{   char ch;
    /*************************/
    FILE fp;
    /*************************/
    fp=fopen("file7.dat","r");
    do
    {   ch=getchar();
        /*************************/
        fprintf(ch,"%c",fp);
    }while(ch!='!');
    fclose(fp);
}
```

(2) 通过函数两次调用，将字符串"Hello"和"World"写入到文件"test"中，成为"Hello World"。

```
#include<stdio.h>
#include<string.h>
void fun(char * fname,char * st)              /*定义函数*/
{   int i,n;
    FILE * fc;                                /*定义文件指针*/
    /*************************/
    fc=fopen("fname","w");
```

```
    n=strlen(st);
    for(i=0;i<n;i++)
/**************************/
        fputs(st[i],fc);
    fclose(fc);
}
void main()
{   fun("test","Hello ");                    /*调用函数*/
    fun("test","World");
}
```

(3) 将文件 test 中的内容全部拷贝到另一个文件 test1 中。

```
#include<stdio.h>
#include<stdlib.h>
void main()
{   FILE *f1,*f2;
    char ch;
/**************************/
    if((f1=fopen("test","w"))==NULL)
    {   printf("Can't open file.\n");  exit(0);  }
/**************************/
     if((f2=fopen("test1","r"))==NULL)
    {   printf("Can't open file.\n");  exit(0);  }
    while(!feof(f1))
/**************************/
    {   ch=fgetc(f2);
/**************************/
        fputc(ch,f1);
    }
    fclose(f1);
    fclose(f2);
}
```

(4) 从文件"test"中读出一个字符,将其加密后写入到"test1"文件中,直至文件末尾为止。加密规则为字符的 ASCII 码值加 1。

```
#include<stdio.h>
void main()
{   char c;
    FILE *fp,*fp1;                          /*定义文件指针*/
/**************************/
    if((fp=fopen("test","w"))==NULL)
    { printf("can't open the file.\n"); return; }
    if((fp1=fopen("test1","w"))==NULL)
    { printf("can't open the file.\n"); return; }
    while(!feof(fp))
```

```
        {   c=fgetc(fp);                        /*从文件中读取字符*/
            /**************************/
            c=c+1;                              /*字符加密*/
            fputc(c,fp1);                       /*写入文件*/
        }
        fclose(fp);
        fclose(fp1);
}
```

(5) 在6～1000之间找出所有的合数,按顺序写入到文件"he.dat"中。所谓"合数"即某数等于其因子之和。例如,28=1+2+4+7+14,则28就是合数。

```
#include<stdio.h>
#include<stdlib.h>
void main()
{   int i,n=6,s;
    FILE * fp;                                  /*定义文件指针*/
    /**************************/
    if((fp=fopen("he.dat","r"))==NULL)
    { printf("can't open the file.\n"); exit(0); }
    while(n<=1000)                              /*找合数*/
    {   s=0;
        for(i=1;i<n;i++)
        /**************************/
            if(n%i!=0)s+=i;
        if(n==s)fprintf(fp,"%5d",n);
        n++;
    }
    fclose(fp);
}
```

(6) 从键盘输入两个学生的数据,写入一个文件中,再读出这两个学生的数据并显示到屏幕上。

```
#include<stdio.h>
#include<stdlib.h>
struct stu                                      /*定义结构*/
{   char name[20];
    int num;
    int age;
    char addr[20];
}a[2],b[2],* pa=a,* pb=b;                       /*定义结构数组和结构指针变量*/
void main()
{   FILE * fp;
    int i;
    /****************************/
    if((fp=fopen("stu.dat","w"))==NULL
```

```
    { printf("can't open the file.\n");exit(0); }
    printf("Input data:\n");
    for(i=0;i<2;i++,pa++)                   /*输入学生信息*/
        scanf("%s%d%d%s",pa->name,&pa->num,&pa->age,pa->addr);
    /*************************/
     pa--;
    fwrite(pa,sizeof(struct stu),2,fp);     /*将学生信息写入文件*/
    rewind(fp);
    fread(pb,sizeof(struct stu),2,fp);      /*读取学生信息*
    printf("name\tnumber\tage\taddr\n");
    /*************************/
    for(i=0;i<2;i++)
        printf("%s %5d%7d %s\n",pb->name,pb->num,pb->age,pb->addr);
    fclose(fp);
}
```

(7) 输入一个字符串并写入到"ou.dat"文件中,把文件中第 0、2、4、…个字符显示输出。

```
#include<stdio.h>
void main()
{   char c;
    FILE * fp;                              /*定义文件指针*/
    /*************************/
    if((fp=fopen("ou.dat","w"))==NULL)
    { printf("can't open the file.\n"); return; }
    while((c=getchar())!='\n')
        fputc(c,fp);                        /*将字符写入文件*/
    rewind(fp);
    c=fgetc(fp);                            /*从文件中读取字符*/
    /*************************/
    while(c!='\0')
    {   putchar(c);
        /*************************/
        fseek(fp,1L,-1);
        c=fgetc(fp);
    }
    printf("\n");
    fclose(fp);
}
```

(8) 下面程序的功能是,显示指定文件的内容,并在显示内容的同时加上行号。从键盘输入指定文件名,若该文件存在并能够打开,将显示其内容,否则将输出"Open file error"。

```
#include<stdio.h>
#include<string.h>
#include<stdlib.h>
void main()
```

```
{   FILE * fp;
    int flag=1,i=0;
    char s[20],filename[20];
    printf("Enter filename:");
    gets(filename);
    if((fp=fopen(filename,"r"))==NULL)
    {   printf("Open file error\n");exit(0);   }
    /***************************/
    while(fgets(fp,20,s)!=NULL)
    /***************************/
    {   if(flag=1)printf("%3d:%s",++i,s);
        else printf("%s",s);
        /***************************/
        if(s[strlen(s)]=='\n')flag=1;
        else flag=0;
    }
    fclose(fp);
}
```

3. 编程题

(1) 已知：$a_0=0, a_1=1, a_2=1, a_n=a_{n-3}+a_{n-2}+a_{n-1}$(当 n 大于 2 时)，求数列 $a_0, a_1, a_2, \cdots, a_{19}$，将数列存储到文件"file1.dat"中，并按照每行 5 个数据输出数列。

(2) a、b、c 为区间[1,100]之间的整数，统计使等式 c/(a*a+b*b)==1 成立的所有解的个数，并将统计数据以格式"%d"写入到文件"file2.dat"中。

(3) 编程将文件"file3.dat"中的数据以行为单位对字符按照由小到大(ASCII 码值)的顺序进行排序，排序后的结果仍按行重新写入到该文件中。原始数据的存放格式是：每行的宽度均小于 80 个字符，可以包含标点符号和空格。例如：

原文：mxdkfi889,j
结果：,889dfijkmx

(4) 将文件"file3.dat"中的字符及其对应的 ASCII 码值显示在屏幕上。

(5) 设有 float 型数组 x[10]和 y[10]，其数组元素 x[i]、y[i]表示平面上某点坐标，统计各点间的最短距离写入到文件"file4.dat"中。从键盘输入各数组元素值。

(6) 对 10 个候选人进行投票选举。现有一个不超过 100 条记录的选票数据文件 in.dat，其数据存放格式为每条记录的长度均为 10 位，第一位表示第一个人的选中情况，第二位表示第二个人的选中情况，以此类推。每一位内容均为字符 1 或 0，1 表示此人被选中，0 表示此人未被选中，若一张选票选中人数小于 5 时则被认为是无效选票。编程统计每个人的得票数，并把每个人的得票数写入到文件"out.dat"中。

(7) 在文件"file5.dat"中写入若干名学生的信息。学生信息包括：学号、3 门课程的成绩、平均分，以格式"%8s%4d%4d%4d%6.2f"写入文件。

(8) 从键盘输入一个学生的信息追加到文件"file5.dat"中。再输入一个学生的学号，修改文件中该名学生的信息，然后将文件中所有学生的信息读出并显示在屏幕上。

(9) 文件"a.dat"和"b.dat"中各存放有一组由小到大排列的数据,现要求将两个文件中的数据合并到一个文件"c.dat"中,仍然按照由小到大的顺序排列。

三、实验步骤

本实验要求掌握文件的基本操作。以一个具体的实例说明文本文件的打开与关闭函数、写入与读出函数的应用。

1. 题目：从键盘输入若干个字符,将其写入到文本文件"disk.txt"中,统计该文件中英文字母、数字、空格和其他字符的个数,要求：

(1) 将统计结果显示输出；

(2) 将文件中的所有字符显示输出；

(3) 将统计结果写入文件 total.txt 中。

2. 算法分析

根据题目要求可知,该问题的处理涉及两个文件的读写。对于 disk.txt 文件要进行先写入后读出的操作,所以打开方式应为"w+",而"total.txt"文件只需要写入,打开方式应为"w"即可。首先需要建立两个文件,然后把若干个字符写入到"disk.txt"中,再从该文件读出显示到屏幕上,并统计其中各种字符的个数,将统计结果写入到"total.txt"中。

3. 根据分析,写出程序代码：

```c
#include<stdio.h>
#include<string.h>
#include<stdlib.h>
void main()
{   char str[80];
    int n,i,y=0,s=0,k=0,q=0;
    FILE *fp;
    printf("请输入字符串：");
    gets(str);                                  /*输入字符串*/
    n=strlen(str);                              /*检测字符串实际长度*/
    if((fp=fopen("disk.txt","w+"))==NULL)
    { printf("Open disk.txt file error\n"); exit(0); }
    fputs(str,fp);                              /*将字符串写入文件*/
    rewind(fp);                                 /*将文件位置移到文件头*/
    fgets(str,n+1,fp);                          /*读文件*/
    fclose(fp);
    puts(str);                                  /*将字符串显示到屏幕上*/
    for(i=0;i<n;i++)                            /*统计英文、数字、空格及其他字符的个数*/
    {   if(str[i]>='a'&&str[i]<='z'||str[i]>='A'&&str[i]<='Z')y++;
        else if(str[i]>='0'&&str[i]<='9')s++;
        else if(str[i]==' ')k++;
        else q++;
    }
    if((fp=fopen("total.txt","w"))==NULL)
```

```
        { printf("Open total.txt file error\n");exit(0);   }
        fprintf(fp,"%d %d %d %d",y,s,k,q);       /*统计结果学到文件中*/
        fclose(fp);
        printf("英文字母:%d,数字:%d,空格:%d,其他:%d\n",y,s,k,q);
}
```

4. 上机调试

将源程序代码录入到 Visual C++ 代码编辑窗口中,然后进行编译、连接,排查并纠正程序中可能存在的语法错误,无误后运行程序。输入字符串"program C!"后输出结果如图 17.1 所示。

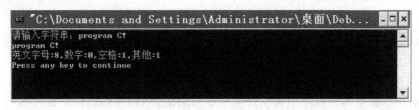

图 17.1 统计文件中字符个数程序运行结果

四、思考题

1. 文件使用完毕后一定要关闭吗？为什么？
2. 文件指针和文件位置指针一样吗？
3. 文件格式读写和块读写要注意些什么？
4. 如何实现文件的随机读写？
5. 在文件的操作中方式"a"和"w"有什么区别？
6. 二进制文件和文本文件有什么区别？

实验十八 综合实验(三)

一、实验目的与要求

1. 巩固前面所学知识点,能够熟练运用,融会贯通。
2. 总结调试程序的经验、技巧。

二、实验内容

1. 填空题

请根据题意在下面各程序中划线处填写适当的语句或表达式,使之能够运行并获得正确的结果。

(1) 数据排序。利用随机数生成函数生成10个整数,通过调用函数sort()将它们由小到大排序,输出时要求奇数在前,偶数在后。

```
#include<stdio.h>
#include<stdlib.h>
#include<time.h>
#define N 10
void Srand(int a[],int n)
{   int i;
    srand(_____);
    for(i=0;i<n;i++)
        a[i]=rand()%100;
}
void main()
{   int i=0,a[N],*p=a;
    void sort(int *p,int n);
    Srand(a,N);
    sort(p,N);
    for(i=0;i<N;i++)
        if(_____) printf("%d ",p[i]);        /*先输出奇数*/
    for(i=0;i<N;i++)
        if(_____) printf("%d ",p[i]);        /*再输出偶数*/
    printf("\n");
}
void sort(int *p,int n)
{   int i,j,k,t;
    for(i=0;i<n-1;i++)
    {   k=i;
```

```
        for(j=i;j<=n-1;j++)
            if(p[k]>p[j])k=j;
        _____;
    }
}
```

(2) 在主函数中建立文件"file.txt",将输入的多条记录写入文件,每条记录包括学号(字符串)和数学成绩(整型),成绩为负数时输入结束。再输入一条记录,调用函数 search()在文件"file.txt"中查找该记录,若存在即输出相应信息;若不存在,则调用 addscore()将该记录添加到文件中,然后输出所有记录。

```
#include<stdio.h>
#include<stdlib.h>
#include<string.h>
#define N 10
struct user
{   char num[10];
    int score;
};
int search(struct user t);
void addscore(struct user t);
void main()
{   int result=0;
    struct user t;
    FILE *fp;
    fp=fopen("file.txt","w");                   /*建立文本文件*/
    if(fp==NULL){ printf("Can't open file!\n"); exit(0); }
    printf("Input record:\n");
    scanf("%s%d",t.num,&t.score);               /*输入一条记录*/
    while(t.score>=0)
    {   fprintf(fp,"%s %d\n",t.num,t.score);    /*写文件*/
        printf("Input next record:\n");
        scanf("%s%d",t.num,&t.score);           /*输入一条记录*/
    }
    fclose(fp);
    printf("Input num and math score:\n");
    scanf("%s%d",t.num,&t.score);               /*输入一条待查找记录*/
    result=search(t);                           /*调用函数在文件中查找该记录*/
    if(result==0)                               /*若没查到,调用函数添加该记录*/
    {   addscore(t);
        printf("Added new records:\n");
    }
    fp=fopen("file.txt",          );
    if(fp==NULL){ printf("Can't open file!\n"); exit(0); }
    while(feof(fp)==0)
    {   fscanf(fp,"%s%d\n",t.num,&t.score);     /*读取文件*/
```

```
            printf("%15s%6d\n",t.num,t.score);          /*显示输出*/
        }
        fclose(fp);
    }
    int search(struct user t)
    {   int i,k=0;
        FILE *fp;
        struct user s[N];
        fp=fopen("file.txt",_____);
        if(fp==NULL){ printf("Can't open file!\n"); exit(0); }
        while(feof(fp)==0)                  /*函数feof返回值为0表示未到文件尾*/
        {   fscanf(fp,"%s %d\n",s[k].num,&s[k].score);
            k++;
        }
        _____;
        for(i=0;i<k;i++)
            if(strcmp(s[i].num,t.num)==0)break;
        if(i!=k)
        {   if(s[i].score==t.score)printf("The record has been in it\n");
            return 1;
        }
        else { printf("The record has been not in it\n");return 0; }
    }
    void addscore(struct user t)
    {   int i,k=0;
        FILE *fp;
        struct user s[N];
        fp=fopen("file.txt",_____);
        if(fp==NULL){ printf("Can't open file!\n"); exit(0); }
        fprintf(fp,"%s %d\n",t.num,t.score);           /*添加新记录*/
        fclose(fp);
    }
```

2. 改错题

下列各个程序中"/*****************/"的下一行中有错误,请仔细阅读程序,根据题意改正程序中的错误,直至调试正确。

(1) 猴子选大王。山上有一群猴子,每个猴子都有编号,编号是 1,2,3,…,m,这群猴子(m 个)按照 1~m 的顺序围坐一圈,从第 1 开始数,每数到第 n 个,该猴子就要离开此圈,这样依次下来,直到圈中只剩下最后一只猴子,则该猴子为大王。建立链表将所有猴子连成一个圆圈。

```
#include<stdio.h>
#include<malloc.h>
struct LNode
{   int  num;
```

```c
        struct LNode * next;
};                                                          /*定义结点*/
struct LNode   * InitList(struct LNode * L,int n)           /*初始化循环链表*/
{    struct LNode * p,* q;
     int i;
     L=(struct LNode * )malloc(sizeof(struct LNode));       /*头结点*/
     L->num=1;                                              /*一号猴子*/
     q=L;
     for(i=2; i<=n;i++)                                     /*从二号猴子开始生成结点*/
     {    p=(struct LNode * )malloc(sizeof(struct LNode));
          p->num=i;
          /*************************/
          p->next=q;
          q=p;
     }
     q->next=L;                                             /*使链表循环起来*/
     return L;
}
void ListDelete_L(struct LNode * L,int n)
{    struct LNode * p,* q;
     int j=1;                                               /*j为计数器,用于数数*/
     p=L;
     while(p->next!=p)                                      /*p->next=p 时是只剩一个结点*/
     {    while(j!=n-1)
          /*当 j=n-1 时应该将该结点的下一个结点删除。当 j!=n-1 时指针向后移,同时计数器
             加一*/
          /*************************/
          {    q=p->next;
               j++;
          }
          q=p->next;                                        /*q 即为被点到的猴子*/
          p->next=p->next->next;                            /*删除 q 结点*/
          free(q);                                          /*释放*/
          j=0;                                              /*计数器清零,重新开始计数*/
     }
     printf("大王为:%d 号猴子。",p->num);                    /*此时的结点就是大王*/
     free(p);
}
int main()
{    struct LNode * L;
     int n,m,e=0;
     printf("请输入猴子个数:");
     scanf("%d",&m);
     printf("请输入 n 值:");
     scanf("%d",&n);
```

```
        if(m<n){printf("m 应该大于 n 请重新输入.\n");return 0;}
        L=InitList(L,m);
        ListDelete_L(L,n);
        printf("\n");
        return 0;
}
```

(2) 文件加密。下面程序在运行时按照规律将文件"file1.txt"中的字符读出，加密后写入另一个文件"file2.txt"中。本程序只对可显示字符进行加密，加密规则为：字符值加 1。

```
#include<stdio.h>
#include<string.h>
#include<ctype.h>
void found()                             /*建立新文件函数*/
{    FILE *fp;
     char p[80];
     int flag=1;
     if((fp=fopen("file1.txt","w"))==NULL)
     {   printf("Can't open file1.\n");flag=0;fclose(fp);   }
     while(flag)
     {   printf("Enter letters:");
         while(strlen(gets(p))>0)
         {   fputs(p,fp);
             fputs("\n",fp);
             fclose(fp);
             flag=0;
         }
     }
}
char New(char ch)                        /*字符加密函数*/
{    if((ch>=32)&&(ch<=122))
         if(ch==122)ch=32;
         else ch=ch+1;
     return ch;
}
void change()                            /*本函数将文件内容读出加密后写入另一文件*/
{    FILE *fp1,*fp2;
     char ch;
     int flag=1;
     /*************************/
     if((fp1=fopen("file1.txt","w"))==NULL)
     {   printf("Can't open file1.\n");flag=0;fclose(fp1); }
     if(flag)
     /*************************/
         if((fp2=fopen("file2.txt","r"))==NULL)
         {   printf("Can't open file2.\n");
```

```
            flag=0;
            fclose(fp1); fclose(fp2);
        }
        while(flag)
        {   /***************************/
            while((ch=fgetc(fp1))==EOF)      /*读文件未到文件尾时循环*/
            {   ch=New(ch);                  /*调用加密函数*/
                fputc(ch,fp2);               /*写入文件*/
            }
            flag=0;
            fclose(fp1); fclose(fp2);
        }
    }
    void main()
    {   char ch;
        int flag=1;
        while(flag)
        {   printf("Input F or f to found a new file.\nC or c to change a file;
            other to exit:");
            ch=toupper(getchar());
            getchar();
            if(ch=='F'||ch=='f')found();
            else if(ch=='C'||ch=='c')change();
            else flag=0;
        }
    }
```

3. 编程题

(1) 编写函数 void fun(int a[],int y,int b[],int *m),找出数组 a 中能被 y 整除并且是奇数的各个整数,按照由小到大的顺序排序存放到数组 b 中,数组 b 的长度由参数 m 返回。主函数为数组 a 输入数据,调用 fun 函数后输出数组 b 的各个元素值。

(2) 编写函数 void fun(int age[],int d[]),统计各年龄段的人数。要求把 0～9 岁年龄段的人数存入 d[0],把 10～19 年龄段的人数存入 d[1],把 20～29 年龄段的人数存入 d[2],依次类推,把 100 岁及以上年龄的人数存入 d[10]。在主函数中通过调用随机函数获得 N 个年龄,并保存在数组 age 中,调用函数 fun 后,输出统计结果。

(3) 利用字符指针编程处理,求所有不超过 200 的 N 值,N 的平方是具有对称性质的回文数。

(4) 八皇后问题。在一个 8×8 的国际象棋盘上,有 8 个皇后,每个皇后占一格;要求棋盘上放上 8 个皇后时不会出现相互"攻击"的现象,即不能有两个皇后在同一行、同一列或对角线上。问共有多少种不同的方法?

(5) 编写程序,将英语名词按照规则由单数变成复数。要求从键盘输入英语规则名词,转换为复数后再输出。已知英语名词单数变复数的规则如下:

① 以辅音字母 y 结尾,将 y 改为 i,再加 es;

② 以 s、x、ch、sh 结尾,加 es;

③ 以元音字母 o 结尾,加 es;

④ 其他情况直接加 s。

(6) 求数字的乘积根,统计 10000 以内的其数字乘积根分别为 1~9 的正整数的个数。所谓数字乘积根,即正整数中非 0 数字的乘积称为该整数的数字乘积,反复取该整数的数字乘积,直到最后剩下一位数字,这个一位数就叫做数字乘积根。例如,1630 的数字乘积为 $1*6*3=18$,18 的数字乘积为 $1*8=8$,因此 8 就是 1630 的数字乘积根。

(7) 要求将不超过 2000 的所有素数从小到大排成第一行,第二行上的每个数都等于它"右肩"上的素数与"左肩"上素数之差。编程求第二行中是否存在这样若干个连续的整数,它们的和正好等于 1898,假如存在的话,又有几种这样的情况? 例如:

第一行:2   3   5   7   11   13   17……1979   1987   1993
第二行: 1   2   2   4   2   4   8   6

(8) 彩票选号。某市体育彩票用整数 1、2、3、…、36 表示 36 种体育运动,一张彩票可选择 7 种运动。编程选择一张彩票的号码,使这张彩票的 7 个号码之和是 105 且相邻两个号码之差按顺序依次是 1、2、3、4、5、6。例如,第一个号码是 1,则后续的号码应是 2、4、7、11、16、22。

(9) 模拟人工洗牌,将洗好的牌分发给 4 个人(提示:利用随机数函数来模拟人工洗牌)。

三、实验步骤

1. 题目:魔术师的猜牌术。魔术师利用一副牌中的十三张黑桃,预先将它们排好后叠放在一起,牌面朝下。魔术师对观众说,我不看牌,只数数就可以猜到每张牌是什么,我数给你们看。魔术师将最上面一张牌数为 1,翻过来正好是黑桃 A,将黑桃 A 放在桌子上,然后按顺序从上到下数 1、2,将第一张放在这叠牌下面,将第二张牌翻过来,正好是黑桃 2,也将它放在桌子上面。第三次数 1、2、3,将前面两张依次放在这叠牌的下面,再翻第三张牌正好是黑桃 3。这样依次进行将 13 张牌全翻出来,准确无误。请问魔术师手中的牌原始次序是怎样排列的?

2. 算法分析

题目已经将魔术师出牌的过程描述的很清楚了,利用倒推的方法,可以很容易的推算出原来牌的排列顺序。在桌子上放十三个空盒子排列成一圈,从 1 开始顺序编号,将黑桃 A 放入 1 号盒中,从下一个空盒开始数数,数到 2 将黑桃 2 放入空盒中,再从下一个空盒数数,数到 3 将黑桃 3 放入空盒中,依次进行,直至将 4、5、…、13 张牌都放入空盒。注意在数数时要跳过非空的盒子,每次数数均从 1 开始。最后牌在盒子中的顺序就是魔术师手中原来牌的顺序。

3. 根据分析,写出程序代码。

定义数组 a,长度为 14,从 a[1] 到 a[13] 按照数数的顺序保存相应的牌,变量 i 表示牌号,i=1 对应黑桃 A,i=2 对应黑桃 2,……,i=13 对应黑桃 K。变量 j 表示数组元素序号,取值为1～13,大于13时使j=1。变量n作为数数时的计数器。

```
#include<stdio.h>
void main()
{   int i,j=1,n,a[14]={0};
    printf("The original order od card is:");
    for(i=1;i<=13;i++)                    /*i代表牌的序号*/
    {   n=1;                              /*空盒计数器*/
        do
        {   if(j>13)j=1;                  /*数到13后再从1开始数*/
            if(a[j])j++;                  /*跳过非空的盒子*/
            else {  if(n==i)a[j]=i; j++;   n++;   }
        }while(n<=i);
    }
    for(i=1;i<=13;i++)
        printf("%d ",a[i]);
    printf("\n");
}
```

4. 上机调试

将源程序代码录入到 Visual C++ 代码编辑窗口中,然后进行编译、连接,排查程序并纠正程序中可能存在的语法错误,无误后运行程序。程序运行结果如图 18.1 所示。

图 18.1　魔术师的猜牌术程序运行结果

第二部分

课 程 设 计

项目一 通 讯 录

一、问题描述与算法分析

随着信息技术的不断发展,计算机的应用也日益广泛,并逐渐深入到人们的日常生活中。通讯录是人们日常生活中经常要使用的通讯管理工具,本软件为一个小型简便通讯录管理程序,可以方便地管理自己的亲朋好友、同学同事、联系接洽用户等的简单资料,为大家提供简便、快捷的管理方法。其中,通讯录数据以文件形式存储在磁盘上,提供记录的录入、删除、修改以及查询等功能。

该通讯录管理系统主要利用单链表实现,主要由四大模块组成:记录输入模块、查询记录模块、更新记录模块、删除记录模块。系统模块结构图如图1.1所示。

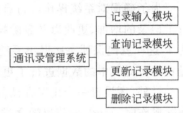

图1.1 系统模块结构图

记录输入模块:该模块主要是将新的记录加入到链表中。在新记录增加操作进行之前,首先需要将已有的通讯录信息从以二进制形式存储的数据文件读入到链表中,接下来再进行记录的增加操作,且通常情况下,记录是增加在链表的末尾。此外,在记录增加的过程,只需要从键盘输入电话号码和联系人姓名,该记录的编号由系统自动增加。作为简易通讯录,目前仅考虑英文姓名或汉语拼音输入。

查询记录模块:该模块主要完成在链表中查找满足相关条件的通讯记录。在该系统中,可以按通讯号码或姓名进行查找。由于指定联系人的联系电话可能有多个,因此,在查询的过程中,会把符合指定条件的记录全部显示出来,以方便接下来的进一步操作。如果没有满足指定条件的记录信息,则显示"未找到"等相关提示信息。

更新记录模块:该模块主要是完成对通讯记录信息的维护。例如,联系人取消了原来的号码而换了新号码,则可以及时对系统中的相应记录进行变更。相对于其他模块来说,该模块的操作较为简单,因此,该模块作为自学部分,要求读者在弄清楚其他模块的具体实现过程之后,自己将该模块的具体代码编写添加到整个系统中。

删除记录模块:该模块主要是删除指定的通讯记录。在本系统中,为了方便记录的删除,首先将满足指定删除条件的记录显示出来,再进一步根据具体的信息删除相应的信息。如果没有满足删除条件的记录,将显示相应的提示信息。

显示记录模块:该模块的主要功能是将所有的通讯录信息按照记录编号由小到大显示输出。

为了实现以上功能,该系统需要定义记录项的格式,其基本属性包括记录编号、姓名、联系电话等。具体如下:

```
struct person
{   int   num;
    char name[15];
```

```
            char tel[15];
    };
```

其中 num 用于表示每条联系人记录的编号，根据记录被加入的顺序，其对应编号依次加 1，数组 name 用于保存联系人的姓名，数组 tel 用于保存联系人的电话，一般情况下不超过 15 位。

链表是实现该通讯录管理系统的主要数据结构，构成链表的结点定义如下：

```
struct node
{   struct person data;                    /*数据域*/
    struct node * next;                    /*指针域*/
};
typedef struct node Node, * Link;
```

为了确保对系统操作进行有效的存储，设置一个全局变量 saveflag，初值为 0。每当进行了记录的增加、更改以及删除操作之后，及时地将变量 saveflag 的值进行修改，重新设置为 1，这样就能保证每次退出系统之前，都会根据变量 saveflag 的值确定在整个系统使用的过程是否对其中的数据进行了更新，以便决定在退出系统之前是否先进行存储操作。

另外，还设置一个用于存储信息记录总数的全局变量 count，每当执行记录的增加或者删除操作时，count 的值也随之进行相应的更改。

该通讯录程序采用了结构化程序设计的思想，程序中除了主函数外，共设计了 8 个函数，具体函数设计如下：

1. menu()

函数原型为 void menu();，该函数用于显示系统在运行过程中所能实现的各个功能构成的主菜单。

2. Disp()

函数原型为 void Disp(Link l);，该函数用于显示单链表中存储的所有联系人记录，其中每一个结点均为 person 结构中定义的内容。

3. printdata()

函数原型为 void printdata(Node * pp);，该函数用于输出链表中某一个结点的数据。

4. Locate()

函数原型为 Node * Locate(Link l,char findmess[]);，该函数用于定位链表中符合要求的结点，并返回指向该结点的指针。

5. Add()

函数原型为 void Add(Link l);，该函数用于往单链表中增加联系人记录对应的结点。

6. Qur()

函数原型为 void Qur(Link l);，该函数用于查找满足条件的联系人记录。

7. Del()

函数原型为 void Del(Link l);，该函数用于删除满足条件的联系人记录对应的结点。

8. Save()

函数原型为 void Save(Link l);，该函数用于将单链表中的数据写入磁盘中的数据文件。

9. main()

函数原型为 void main();，该函数用于实现对整个程序的运行控制,以及相关功能模块的调用。它先以可读写的方式打开数据文件"c:\\phonebook,"若文件不存在,则新建该文件。当打开文件操作成功后,从文件中一次读出一条记录,添加到新建的单链表中,然后执行显示主菜单并进入主循环操作,接下来便可以根据键值的不同执行不同的操作。

二、难点提示

C 程序设计语言是计算机科学与技术专业的基础课程,在学习该课程时,学生往往还没有学习过数据库原理。因此,在该项目设计的过程中,如何有效的保存通讯记录信息是系统设计的难点之一。在此,是通过一个专门的文本文件来记录所有的通讯录信息的。这样,在系统设计的过程中,就要涉及有关文本文件的读、写等操作。

三、部分代码

以下代码包含除实现修改功能以外的所有功能,主要有增加记录、查询记录、删除记录、显示记录、保存记录。为了保证程序的正确运行,对主菜单也进行了相应地调整。读者可以在正确理解各个模块的具体实现的基础上,对程序进行修改,使之也能够实现记录的更新操作。相对于其他操作来说,更新操作是最为简单的。

1. 主函数 main():主要实现对整个程序的运行控制,以及相关功能模块的调用。具体代码如下:

```c
void main()
{   Link l;                                 /*定义链表*/
    FILE * fp;                              /*文件指针*/
    int select;                             /*保存选择结果变量*/
    char ch;                                /*保存(y,Y,n,N)*/
    Node * p,* r;                           /*定义记录指针变量*/
    l=(Node *)malloc(sizeof(Node));
    if(!l)
    {   printf("allocate memory failure!\n");  /*如没申请到内存,输出提示信息*/
        return;                             /*返回主界面*/
    }
    l->next=NULL;
    r=l;
    fp=fopen("c:\\phonebook","ab+");        /*以追加方式打开一个二进制文件*/
    if (fp==NULL)
    {   printf("\n   =====can not open file!=====\n");
        exit(0);
    }
    while (!feof(fp))                       /*统计联系人记录的个数*/
    {   p=(Node *)malloc(sizeof(Node));
```

```c
        if(!p)
        {   printf(" memory malloc failure!\n");
            return;
        }
        if(fread(p,sizeof(Node),1,fp)==1)            /*一次从文件中读取一条记录*/
        {   p->next=NULL;
            if(count==0) { l->next=p;    r=l; }
            r->next=p;
            r=p;                                      /*r指针向后移动一个位置*/
            count++;
        }
    }
    fclose(fp);                                       /*关闭文件*/
    system("cls");
    printf("open file sucess,the total records number is :%d\n",count);
    menu();
    while(1)
    {   system("cls");
        menu();
        p=r;
        printf("Please Enter your choice 0-4:");      /*显示提示信息*/
        scanf("%d",&select);
        switch (select)
        {   case 0:
            {   if(saveflag==1)                        /*若对链表的数据有修改且未存盘,则此标志为1*/
                {   getchar();
                    printf("\n      =====Whether save the modified record to file?(y/n))?");
                    scanf("%c",&ch);
                    if(ch=='y'||ch=='Y')Save(l);
                }
                printf("   =====thank you for useness!");
                exit(0);
            }
            case 1:Disp(l);   break;                   /*显示联系记录*/
            case 2:Add(l); break;                      /*增加联系人记录*/
            case 3:getchar(); Del(l);  break;          /*删除联系人记录*/
            case 4:Qur(l); break;                      /*查询联系人记录*/
            default: getchar(); printf("\n       **********ERROR:input has wrong!
            press any key to continue**********\n"); getchar(); getchar();
            getchar();                                 /*按键有误,必须为数值0~4*/
        }
    }
}
```

2. 主菜单界面：用户进入联系人信息管理系统时，首先显示主菜单，提示用户进行选择，完成相应任务。具体代码如下：

```
void menu()                        /*主菜单*/
{   system("cls");                 /*调用 DOS 命令,清屏.与 clrscr()功能相同*/
    cprintf("       the information management system of phonebook\n\n\n");
    cprintf("    ************************MENU************************\n");
    cprintf("              1   display  record \n");
    cprintf("              2   insert   record \n");
    cprintf("              3   delete   record \n");
    cprintf("              4   search   record\n");
    cprintf("              0   quit     system \n");
    cprintf("    ****************************************************\n");
                                   /*cprintf()送格式化输出至文本窗口屏幕中*/
}
```

3. 表格形式显示记录：由于记录显示操作经常进行，所以将这部分由独立的函数来实现，减少代码的重复。它将显示单链表 l 中存储的联系人记录，内容为 person 结构中定义的内容。具体代码如下：

```
void Disp(Link l)                  /*显示单链 l 中存储的联系人记录*/
{   Node *p;
    p=l->next;                     /*l 存储的是单链表中头结点的指针,该头结点没有存储
                                     联系人信息,指针域指向的后继结点才有联系人信息*/
    if (!p)                        /*p==NULL,NULL 在 stdlib 中定义为 0*/
    {   getchar();
        printf("\n      ======Not person record!  Press any key to continue!\n");
        getchar();
        return;
    }
    while (p)                      /*逐条输出链表中存储的联系人信息*/
    {   printdata(p);
        p=p->next;
    }
    printf("\n\n      ==========Press any key to continue!\n");
    getchar();
    getchar();
}
```

4. 记录查找定位：用户进入信息管理系统时，对某个联系人的记录进行处理前，需要按照条件找到这条记录。具体代码如下：

```
Node * Locate(Link l,char findmess[])             /*按姓名查询*/
{   Node *r;
    r=l->next;
    while (r)
```

```
    {   if (strcmp(r->data.name,findmess)==0)      /*若找到 findmess 值的联系人*/
            return r;
        r=r->next;
    }
    return 0;                                       /*若未找到,返回一个空指针*/
}
```

5. 格式化输入数据:由于该系统中,要求用户输入的只有字符型和数值两种类型,所以设计以下函数来单独处理,并对输出的数据进行检验。具体代码如下:

```
void stringinput(char *t, int lens, char *notice)
{   char n[255];
    do
    {   printf(notice);                             /*显示提示信息*/
        scanf("%s",n);                              /*输入字符串*/
        if (strlen(n)>lens)                         /*进行长度校验,超过 lens 值重新输入*/
            printf("\n\n    exceed the required length!\n");
    } while (strlen(n)>lens);
    strcpy(t,n);                                    /*将输入的字符串拷贝字符串 t 中*/
}
```

6. 增加联系人记录:若数据文件为空,则从单链表的头部开始增加联系人记录结点;否则,将此联系人记录结点添加在单链表的尾部。具体代码如下:

```
void Add(Link l)
{   Node *p,*r,*s;                                  /*实现添加操作的临时的结构指针变量*/
    char ch,flag=0,tel[15];
    r=l;
    s=l->next;
    while(r->next!=NULL)
        r=r->next;                                  /*将指针移至链表最末尾,准备添加记录*/
    while (1)        /*一次可输入多条记录,直至输入手机号为 0 的记录结束添加操作*/
    {   while (1)    /*输入手机号,保证该手机号未被使用,若手机号为 0,则退出添加*/
        {   stringinput(tel,15,"\n    input tel number(press '0' return menu):");
                     /*格式化输入手机号并检验*/
            flag=0;
            if (strcmp(tel,"0")==0)return;          /*输入为 0,则退出添加操作,返回主界面*/
            s=l->next;
            while (s)    /*查询该手机号是否已存在,若存在则要求重新输入一个手机号*/
            {   if (strcmp(s->data.tel,tel)==0)
                {   flag=1;
                    break;
                }
                s=s->next;
            }
            if (flag==1)                            /*提示用户是否重新输入*/
```

```
            {   getchar();
                printf("\n      =====The tel number %s is existing,try again?
                 (y/n):",tel);
                scanf("%c",&ch);
                if (ch=='y'||ch=='Y')  continue;
                else return;
            }
            else break;
        }
        p=(Node *)malloc(sizeof(Node));      /*申请内存空间*/
        if (!p)
        {   printf("\n     allocate memory failure\n");
            return;
        }
        strcpy(p->data.tel,tel);             /*将字符串tel拷贝到p->data.tel中*/
        stringinput(p->data.name,15,"\n      Input name:");     /*输入姓名*/
        count++;
        p->data.num=count;
        p->next=NULL;                        /*表明这是链表的尾部结点*/
        r->next=p;                           /*将新建的结点加入链表尾部中*/
        r=p;
        saveflag=1;
        printf("\n      ========Insert a recdord sucess!\n\n");
    }
}
```

7. 查询联系人记录：当用户执行此查询任务时，系统会提示用户进行查询字段的选择，即按手机号或姓名进行查询。若记录存在，则会输出此联系人记录的信息。具体代码如下：

```
void Qur(Link l)                    /*按手机号或姓名查找满足条件的联系人记录*/
{   char searchinput[15];           /*保存用户输入的查询内容*/
    Node *p;
    if (!l->next)                   /*若链表为空*/
    {   getchar();
        printf("\n      ======No person record!Press any key to continue!\n");
        getchar();
        return;
    }
    stringinput(searchinput,15,"\n    input the existing person number:");
    p=Locate(l,searchinput);
                        /*在l中查找联系人为searchinput的结点,并返回其指针*/
    if (p)
    {   printdata(p);
        while(p->next!=NULL)  /*在l中查找姓名为searchinput的结点,并返回其指针*/
```

```
            {   p=p->next;
                if (strcmp(p->data.name,searchinput)==0)   printdata(p);
            }
            printf("\n    press any key to return!\n");
            getchar();
        }
        else
        {   getchar();
            printf("\n    ======Not find this person!\n\n    press any key to
            return!\n");
            getchar();
        }
    }
```

8. 删除联系人记录：在删除操作中，系统会按用户要求先找到联系人记录的结点，然后从单链表中删除该结点。具体代码如下：

```
    void Del(Link l)                /*删除满足条件的联系人记录的结点*/
    {   Node  *p,*r,*s;
        char namemess[15],telmess[15];
        int flag=0;
        if(!l->next)
        {   system("cls");
            printf("\n    ======No person record!\n");
            return;
        }
        stringinput(namemess,15,"\n    input the existing person name:");
        s=Locate(l,namemess);
        if(s)                        /*显示输出联系人namemess的所有电话号码*/
        {   printf("\n    Here are telphones of %s that you    want to delete probably:
            \n\n",namemess);
            printdata(s);
            flag=1;
            while(s->next!=NULL)
            {   s=s->next;
                if (strcmp(s->data.name,namemess)==0)   printdata(s);
            }
        }
        if(flag==1)                  /*如果存在联系人namemess,则将其指定号码删除*/
        {   stringinput(telmess,15,"\n    input the existing telphone that you want to
            delete:");
            p=l->next;
            r=l;
            while(1)
            {   if(strcmp(p->data.tel,telmess)==0)
```

```
            {   if(p->next!=NULL)
                {   r->next=p->next;
                    free(p);
                    p=r->next;
                    do            /*将后面联系人的编号 num 依次减 1*/
                    {   p->data.num=p->data.num-1;
                        p=p->next;
                    }while  (p);
                    p=r->next;
                }
                else
                {   r->next=NULL;  free(p);   }
                count--;
                saveflag=1;
                flag=0;
                break;
            }
            else
            {   if(p->next!=NULL){ r=r->next;p=p->next;  }
                else break;
            }
        }
        if(flag==1)
        {   printf("\n\n    =====the telephone that you wanted do not exit! Press
            any key to continue! \n");
            getchar();
        }
        else
        {   printf("\n\n    ========delete success!   Press any key to continue
            \n");
            getchar();
        }
    }
    else                 /*如果不存在联系人 namemess 的电话号码,则输出提示信息*/
    {   printf("\n    =======Not find this person! And you don't detele any
        information!\n\n    Press any key to return!\n");
        getchar();
    }
}
```

9. 存储联系人记录：在存储联系人记录操作中，系统会将单链表中数据写入至磁盘中的数据文件，若用户对数据有修改而没有专门进行此操作，那么在退出系统时，系统会提示用户是否存盘。具体代码如下：

```
void Save(Link l)              /*用于将单链表l中的数据写入磁盘中的数据文件*/
{   FILE *fp;
    Node *p;
    fp=fopen("c:\\phonebook","wb");         /*以只写方式打开二进制文件*/
    if (fp==NULL)                            /*打开文件失败*/
    {   printf("\n       =======open file error!\n");
        return;
    }
    p=l->next;
    while(p)
    {   if(fwrite(p,sizeof(Node),1,fp)==1)   /*每次写一个结点信息至文件*/
        {  p=p->next;  count++;   }
         else break;
    }
    if(count>0)
    {   printf("\n\n      =========Save file complete,total save's record
        number is:%d\n",count);
        saveflag=0;
    }
    else
       printf("     the current link is empty,no person record is saved!\n");
    fclose(fp);                              /*关闭文件*/
}
```

四、运行结果示例

通讯录管理系统的界面设计主要遵循方便易用、界面友好的原则,具体如下:
1. 主界面

运行程序,首先打开系统的主界面,如图 1.2 所示。在此,可以根据实际的需要选择相应的操作,其中选择 0~4,分别对应:退出系统、浏览记录、插入记录、删除记录和查询记录等操作。

图 1.2 系统主界面

2. 显示记录界面

用户在进行增加、删除等操作之前，一般都需要对已有的信息记录进行浏览。浏览信息记录的操作相对其他操作来说较简单，用户只需要在主界面中输入数字"1"后按下 Enter 键，所有的信息记录就会显示出来，如图 1.3 所示。

图 1.3　浏览信息记录界面

3. 插入记录界面

在主界面中输入数字"2"，并按下 Enter 键，接下来就可以实现增加记录的操作，如图 1.4 所示。首先提示输入联系人的电话号码，如果输入的号码不存在，则可以继续输入联系人的姓名，如图 1.5 所示。输入姓名后，按下回车键，系统首先提示该条信息添加成功，同时提示可以继续输入下一条信息，如图 1.6 所示。如果此时不想再继续输入，可以直接输入数字"0"结束增加记录的操作，回到主界面。如果输入的号码已经存在，则系统会给出提示信息，如图 1.7 所示。用户可以根据实际情况选择"y"或"n"。

图 1.4　选择数字"2"之后出现的画面

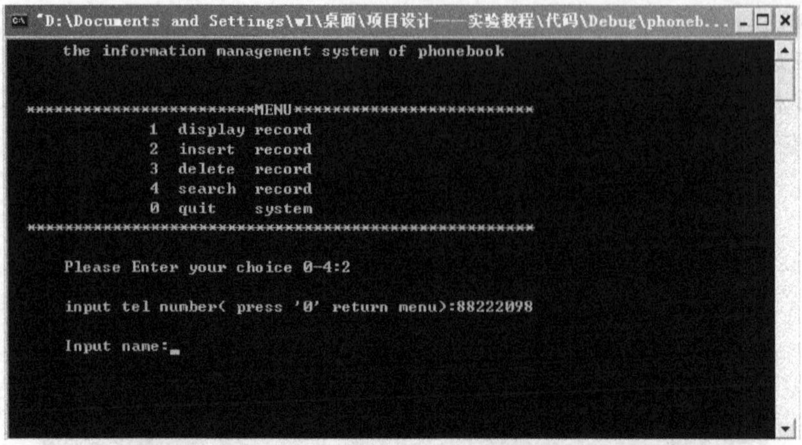

图1.5 提示输入联系人姓名的界面

图1.6 显示添加信息成功以及开始输入下一条信息的界面

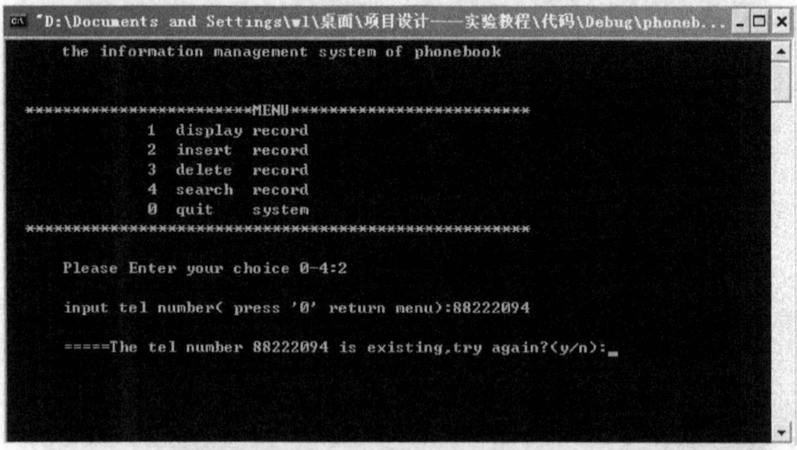

图1.7 显示添加的电话号码已经存在的界面

4. 删除记录界面

在实际使用的过程中,用户可以根据实际需要删除一些不需要的记录信息。具体如下:首先在主界面中输入数字"3",按下 Enter 键后进入信息记录删除界面,如图 1.8 所示。根据提示信息输入联系人的姓名,由于同一个联系人可能会有多个联系电话,接下来会把有关该联系人的所有联系电话都显示出来,如图 1.9 所示。根据实际情况,选择具体的电话号码进行删除操作。删除操作完成后,系统会自动给出"操作完成"提示信息,如图 1.10 所示。

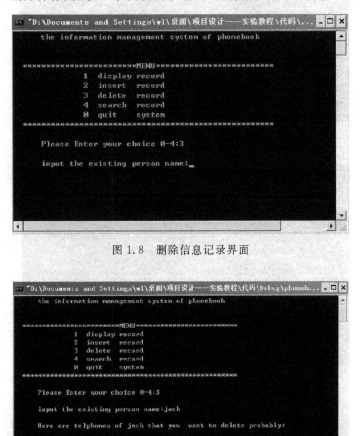

图 1.8 删除信息记录界面

图 1.9 显示联系人所有的联系电话并提示输入具体要删除的信息界面

5. 查找记录界面

信息的查找也是系统的主要功能之一。在主界面中输入数字"4"后按下回车键,即进入信息查询界面,如图 1.11 所示。根据提示信息输入被查找人的姓名后,与该姓名相关的电话信息就显示出来,如图 1.12 所示。

6. 退出系统界面

需要退出系统,只需要在主界面中输入数字"0"后按下回车键即可。

图 1.10 删除信息成功界面

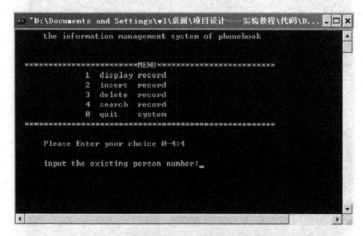

图 1.11 查询界面

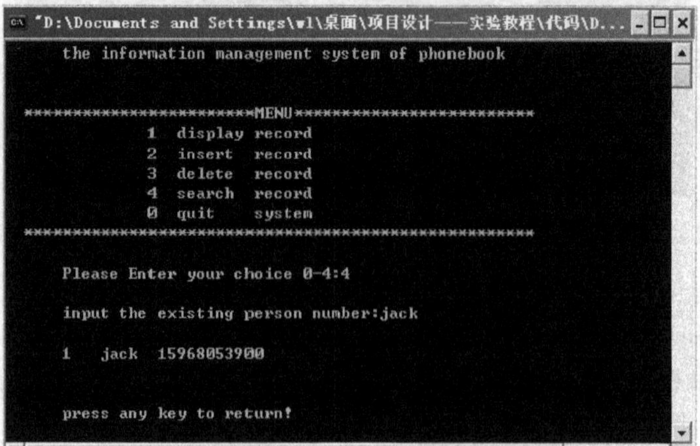

图 1.12 显示相关联系人的电话信息界面

项目二　五子棋游戏

一、问题描述与算法分析

五子棋是起源于中国古代的传统黑白棋种之一。五子棋不仅能增强思维能力，提高智力，而且富含哲理，有助于修身养性。五子棋既有现代休闲的明显特征"短、平、快"，又有古典哲学的高深学问"阴阳易理"；它既有简单易学的特性，为人民群众所喜闻乐见，又有深奥的技巧和高水平的国际性比赛；它的棋文化源远流长，具有东方的神秘和西方的直观；既有"场"的概念，亦有"点"的连接。它是中西文化的交流点，是古今哲理的结晶。

五子棋游戏是常见的经典小游戏，其规则简单。具体如下：选定一方先下，然后黑白双方依次轮流下子。棋盘上形成横向、竖向、斜向的连续的相同颜色的五个棋子称为"五连"。黑白双方先在棋盘上形成五连的一方为胜。若对局双方均认为不可能形成五连或是剩余棋盘空间已不足以形成五连则为和棋。

游戏中的棋盘用一个 18×18 的二维数组表示，具体如下：

int box[N][N];

其中数组的每一个元素对应棋盘上的一个交叉点，用"0"表示空位，用"1"代表玩家1的子，"2"代表玩家2的子。整个游戏主要包括游戏初始化模块、游戏结束清理模块、主循环控制模块、键盘处理模块和胜负判别模块。系统总体结构图如图2.1所示。

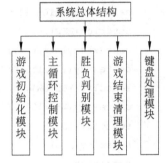

图 2.1　系统模块结构图

该游戏程序采用了结构化程序设计的思想，程序中除了主函数外，共设计了7个函数，具体函数设计如下：

1. init()

函数原型为 void init();，该函数用于初始化棋盘数组以及绘制棋盘。

2. draw_circle()

函数原型为 void draw_circle(int x, int y, int color);，用于在指定的坐标位置(x,y)用指定颜色绘制一个棋子。

3. change()

函数原型为 void change();，为了标识下棋过程中的某一时刻轮流到了哪一方，在程序中用一个全局变量 flag 进行标识。该函数就是用于改变标志状态 flag 的。

4. judgewho()

函数原型为 void judgewho(int x,int y);，该函数用于判断该哪方下子。

5. judgekey()

函数原型为 void judgekey();，该函数用于判断输入的按键，并做出处理。

6. judgeresult

函数原型为 int judgeresult(int x,int y);,该函数用于判断输赢结果。

7. re_draw_box()

函数原型为 void re_draw_box();,该函数用于移动棋子后恢复棋盘原貌。

8. disp(int flag)

函数原型为 void disp(int flag);,该函数展示游戏一局结束时的界面。

9. void inf()

函数原型为 void inf();,该函数展示游戏开始时的界面。

二、难点提示

胜负判别是五子棋开发过程中的难点之一,双方每下一个子时都需要进行胜负判别。在此,对于胜负的判别就是统计以所下子为中心,考察沿水平、垂直、正对角线、反对角线上,以此相连的同种颜色的棋子的个数。如果其中有一条线上构成了连续的同种颜色的 5 个棋子,则表示该盘棋局已经分出胜负。详见部分代码中的 int judgeresult(int x,int y); 函数。

三、部分代码

1. 函数 inf()   /*游戏开始时的界面*/

```
void inf();
{   setbkcolor(LIGHTBLUE);
    setcolor(RED);
    settextstyle(1,0,8);
    outtextxy(159,150,"WELCOME");
    setcolor(YELLOW);
    outtextxy(154,150,"WELCOME");
    setcolor(15);
    setcolor(RED);
    settextstyle(3,0,1);
    outtextxy(100,350,"NOTE:DON'T PRESS SPACE OUTSIDE OF THE CHESSBOARD!");
    setcolor(WHITE);
    outtextxy(350,400,"PRESS ANY KEY TO START");
    getch();
    getch();
    cleardevice();
}
```

2. 函数 disp()   /*游戏一局结束时的界面*/

```
void disp(int flag)
{   char ch;
```

```c
    cleardevice();
    setbkcolor(LIGHTBLUE);
    setcolor(RED);
    settextstyle(1,0,6);
    if(flag==1)  outtextxy(190,180,"WHITE WIN");
    else    outtextxy(192,180,"RED WIN");
    settextstyle(1,0,3);
    setcolor(WHITE);
    outtextxy(350,300,"REPLAY?    [Y/N]");
    while (1)
    {    while(bioskey(1)==0);
         ch=bioskey(0);
         if (ch=='N'||ch=='n')   exit(0);
         else  if(ch=='Y'||ch=='y'){  cleardevice();  break;   }
    }
}
```

3. 函数 re_draw_box()   /* 用于移动棋子后恢复棋盘原貌 */

```c
void re_draw_box()
{   int x1,x2,y1,y2;
    setbkcolor(LIGHTBLUE);
    setcolor(YELLOW);
    gotoxy(7,2);
    for(x1=1,y1=1,y2=18;x1<=18;x1++)
        line((x1+JZ) * BILI,(y1+JS) * BILI,(x1+JZ) * BILI,(y2+JS) * BILI);
    for(x1=1,y1=1,x2=18;y1<=18;y1++)
        line((x1+JZ) * BILI,(y1+JS) * BILI,(x2+JZ) * BILI,(y1+JS) * BILI);
}
```

4. 函数 init()   /* 初始化数组以及绘制棋盘 */

```c
void init();
{   int x1,x2,y1,y2;
    setbkcolor(LIGHTBLUE);
    setcolor(YELLOW);
    gotoxy(7,2);
    printf("LEFT,RIGHT,UP,DOWN KEY TO MOVE,SPACE TO PUT,ESC QUIT");
    for(x1=1,y1=1,y2=18;x1<=18;x1++)
        line((x1+JZ) * BILI,(y1+JS) * BILI,(x1+JZ) * BILI,(y2+JS) * BILI);
    for(x1=1,y1=1,x2=18;y1<=18;y1++)
        line((x1+JZ) * BILI,(y1+JS) * BILI,(x2+JZ) * BILI,(y1+JS) * BILI);
    for (x1=1;x1<=18;x1++)
        for (y1=1;y1<=18;y1++)
            box[x1][y1]=0;
}
```

5. 函数 draw_cicle()   /* 用指定颜色在(x,y)坐标处绘制一个棋子 */

```c
void draw_cicle(int x, int y, int color)
{   setcolor(color);
    x= (x+JZ) * BILI;
    y= (y+JS) * BILI;
    setfillstyle(1,color);
    fillellipse(x,y,8,8);
}
```

6. 函数 judgekey()   /* 用于判断输入的按键,并做出处理 */

```c
void judgekey()
{   int i,j;
    switch(key)
    {   case LEFT:
            if(step_x-1<0)break;
            else
            {   for(i=step_x-1,j=step_y;i>=1;i--)
                    if(box[i][j]==0)
                    {   draw_circle(step_x,step_y,LIGHTBLUE);
                        re_draw_box(step_x,step_y,LIGHTBLUE);
                        break;
                    }
                if(i<1)break;
                step_x=i;
                judgewho(step_x,step_y);
                break;
            }
        case RIGHT:
            if(step_x+1>18)break;
            else
            {   for(i=step_x+1,j=step_y;i<=18;i++)
                    if(box[i][j]==0)
                    {   draw_circle(step_x,step_y,LIGHTBLUE);
                        re_draw_box(step_x,step_y,LIGHTBLUE);
                        break;
                    }
                if(i>18) break;
                step_x=i;
                judgewho(step_x,step_y);
                break;
            }
        case DOWN:
            if(step_y+1>18) break;
```

```c
            else
            {   for(i=step_x,j=step_y+1;j<=18;j++)
                    if(box[i][j]==0)
                    {   draw_circle(step_x,step_y,LIGHTBLUE);
                        re_draw_box(step_x,step_y,LIGHTBLUE);
                        break;
                    }
                if(j>18) break;
                step_y=j;
                judgewho(step_x,step_y);
                break;
            }
        case UP:
            if(step_y-1<0) break;
            else
            {   for(i=step_x,j=step_y-1;j>=1;j--)
                    if(box[i][j]==0)
                    {   draw_circle(step_x,step_y,LIGHTBLUE);
                        re_draw_box(step_x,step_y,LIGHTBLUE);
                        break;
                    }
                if(j<1) break;
                step_y=j;
                judgewho(step_x,step_y);
                break;
            }
        case ESC:   break;
        case SPACE:
            if(step_x>=1&&step_x&&step_y>=1&&step_y<=18)
            {   if(box[step_x][step_y]==0)
                {   box[step_x][step_y]=flag;
                    if(judgeresult(step_x,step_y)==1)
                    {   sound(1000);
                        delay(1000);
                        nosound();
                        disp(flag);
                        re=1;
                    }
                    change();
                    break;
                }
            }
            else   break;
    }
}
```

7. 函数 change()　　　　　　　/*用于改变标志状态*/

```
void change()
{   if(flag==1)flag=2;
    else   flag=1;
}
```

8. 函数 judgewho()　　　　　　/*用于判断是哪方的棋子*/

```
void judgewho(int x,int y)
{   if(flag==1)
    {   draw_circle(x,y,15);
        gotoxy(60,15);
        printf("WHITE MOVE");
    }
    if(flag==2)
    {   draw_circle(x,y,4);
        gotoxy(60,15);
        printf("RED MOVE");
    }
}
```

9. 函数 judgeresult()　　　　　/*用于判断输赢结果*/

```
int judgeresult(int x,int y)
{   int j,k,n1,n2;
    while(1)
    {   n1=0;
        n2=0;
        /*水平向左数*/
        for(j=x,k=y;j>=1;j--)
        {   if(box[j][k]==flag)n1++;
            else   break;
        }
        /*水平向右数*/
        for(j=x,k=y;j<=18;j++)
        {   if(box[j][k]==flag)n2++;
            else    break;
        }
        if(n1+n2-1>=5) return(1);
        /*垂直向上数*/
        n1=0;
        n2=0;
        for(j=x,k=y;k>=1;k--)
        {   if(box[j][k]==flag) n1++;
            else   break;
        }
```

```
        /*垂直向下数*/
        for(j=x,k=y;k<=18;k++)
        {   if(box[j][k]==flag)n2++;
            else    break;
        }
        if(n1+n2-1>=5) return(1);
        /*向左上方数*/
        n1=0;
        n2=0;
        for(j=x,k=y;j>=1||k>=1;j--,k--)
        {   if(box[j][k]==flag)n1++;
            else    break;
        }
        /*向右下方数*/
        for(j=x,k=y;j<=18||k<=18;j++,k++)
        {   if(box[j][k]==flag)n2++;
            else    break;
        }
        if(n1+n2-1>=5) return(1);
        /*向右上方数*/
        n1=0;
        n2=0;
        for(j=x,k=y;j<=18||k>=1;j++,k--)
        {   if(box[j][k]==flag) n1++;
            else break;
        }
        /*向左下方数*/
        for(j=x,k=y;j>=1||k<=18;j--,k++)
        {   if(box[j][k]==flag) n2++;
            else    break;
        }
        if(n1+n2-1>=5)return(1);
        return(0);
    }
}
```

10. 主函数 main()   /*程序的主控制模块*/

```
void main()
{   int gdriver=VGA,gmode=VGAHI;
    clrscr();
    initgrapg(&gdriver,&gmode,"C:\\TC");
    inf();
    while(key!=ESC)
    {   re=0;
        flag=1;
```

```
        init();
        do
        {   step_x=0;
            step_y=0;
            judgewho(step_x-1,step_y-1);
            do
            {   while(bioskey(1)==0);
                key=boiskey(0);
                judgekey();
            }
            while(key!=SPACE && key!=ESC);
        }
        while(key!=ESC&&key!=1);
    }
    closegraph();
}
```

四、运行结果示例

1. 游戏打开时的状态

运行五子棋游戏,进入如图 2.2 所示的界面。按任意键开始游戏,进入游戏的初始状态,如图 2.3 所示。

图 2.2 运行游戏时开始画面

2. 游戏的初始状态

开始游戏后,进入游戏的初始状态,如图 2.3 所示。通过按键盘上的↑、↓、←、→箭头键可以移动棋子的位置,最后通过按下 Space(空格)键确定棋子的位置。

3. 游戏进行状态

一方下过一个棋子之后,就轮到另一方下一颗棋子,如图 2.4 和图 2.5 所示。图 2.4 是轮到白方下棋,图 2.5 是轮到红方下棋。在双方交战的过程中,都是通过键盘上的↑、↓、←、→箭头键来移动棋子的,并按下 Space(空格)键确定最终棋子的位置。每一次按下 Space(空格)键确定最终棋子的位置后,系统就自动进行胜负评判。如果能确定双方的胜负,游戏就进入一局结束状态,如图 2.6 所示。

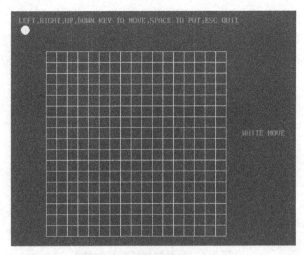

图 2.3　游戏开始时的状态

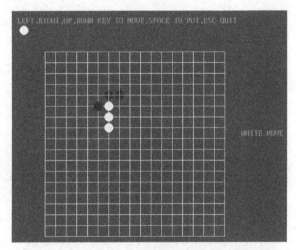

图 2.4　轮到白方下子

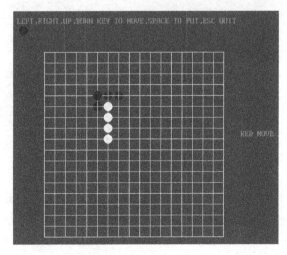

图 2.5　轮到红方下子

4. 游戏结束状态

一局游戏结束,系统会根据对整个棋面的评判结果给出相应的信息,同时用户可以根据实际决定是否继续游戏,如图2.6所示。用户按下"Y"则继续下一局游戏,按下"N"则结束游戏。

图2.6 一局游戏结束后的状态

项目三　英汉电子词典

英汉电子词典作为一个常用的学习工具,是经常要使用的,本系统要求能完成一个简单的英汉电子词典的功能,实现单词的查找、增加、删除、显示和排序工作。

1. 算法分析

(1) 在计算机中建立有限规模的英汉电子词典,该词典的内容以文件作为数据库来存储,利用本系统实现对文件中单词词条的查找、增加、删除、显示和排序工作。

(2) 词典的内容为,每行对应一个词条,每个词条由两个字符串组成,两个字符串之间用若干空格符分开,前一个字符串是英文单词字符串,后一个字符串是该英文单词的中文释义字符串,多个释义字符串之间可用分号作为分隔符(不要有空格)。

(3) 对英文单词和中文释义字符串长度的限定分别为不多于 20 个字符与 80 个字符,单词词条最多为 200 条。

(4) 采用菜单工作方式,在一个操作执行后,询问是否继续执行该操作,若输入 Y 或 y,则重复执行该操作,不退回菜单;若输入其他信息,则回到菜单等待另一次选择;当选择退出操作时,退出系统,程序运行结束。

根据以上算法分析,该系统的主框架功能结构如图 3.1 所示。

图 3.1　程序主框架功能结构图

程序流程图如图 3.2 所示。

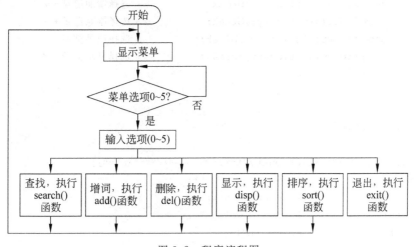

图 3.2　程序流程图

2. 难点提示

（1）程序需要定义一个全局结构体类型 struct dict，它有两个成员 word（英文单词）、chinese（中文释义），此结构体类型在每个函数中都可使用；结构体数组 dictionary 在 main() 中定义，此数组作为各函数的参数调用。

（2）程序除 main() 函数外，另外有 5 个函数，函数 search() 用于查找词条，在此函数中先从键盘输入要查找的单词，打开数据库文件，把文件中的内容读入结构体数组，用循环语句从结构体数组中调出符合条件的单词及其释义；add() 函数用于增加记录，在此函数中先从键盘输入要增加的记录数，然后增加相应数目的词条，增加完后根据实际需要选择是否追加到数据库文件；函数 del() 用于删除记录，在此函数中先从键盘读入要删除的单词，打开数据库文件，把文件中的内容读入结构体数组，再将结构体数组中的内容写入文件，其中符合删除条件的词条就不再写入文件，以此方法来删除相应的词条；函数 sort() 用于对数据文件中的词条按英文单词的字母进行排序，在此函数中先打开数据库文件，把文件中的内容读入结构体数组中，再用选择排序法对数据组中的数据按照英文单词的字母进行排序，再将排序好的结构体数组中的数据写入数据库文件；函数 disp() 用于显示记录，在此函数中先打开数据库文件，把文件中的内容读入结构体数组，然后用循环方法将数组中的内容输出到屏幕上。

（3）为了程序有个友好的用户界面，可以自定义一个菜单函数 menu_select() 负责显示菜单和传递用户的选择，在 main() 函数中用一个无限循环 for(;;) 或者 while(1)，等待用户选择菜单。

3. 部分代码

这里只给出主函数和菜单函数的代码，其他函数请读者自己根据以上提示信息写出。

```c
void main()
{   struct  dict dictionary[M];                  /*在各函数中使用*/
    for(;;)
    {   switch(menu_select())
       { case 0:exit(0);                          /*选择退出*/
         case 1:search(dictionary);break;         /*选择查找记录*/
         case 2:add(dictionary);break;            /*选择增加记录*/
         case 3:del(dictionary);break;            /*选择删除记录*/
         case 4:sort(dictionary);break;           /*选择记录排序*/
         case 5:disp(dictionary);break;           /*选择显示记录*/
       }
    }
}
int menu_select()
{   int c;
    printf("\n*********Dictionary Menu***************\n\n");
    printf("            1. search word\n");
    printf("            2. add word\n");
    printf("            3. delete word\n");
    printf("            4. sort\n");
    printf("            5. display\n");
    printf("            0. exit\n\n");
```

```
    printf("*****************************************************\n");
    do{
        printf("\n请输入您的选择项(0~5):");
        scanf("%d",&c);
    }while(c<0||c>5);
    return c;
}
```

4. 运行结果示例

（1）程序开始运行时，显示菜单如图3.3所示。

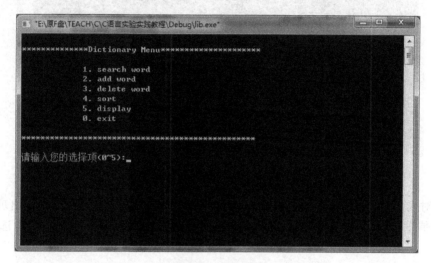

图3.3　显示菜单

（2）输入菜单选项2 add word后，先输入要录入的词条数目，这里输入3个。当输入完3个词条后，系统会问是否要保存，输入1为保存，输入0放弃保存。运行结果如图3.4所示。

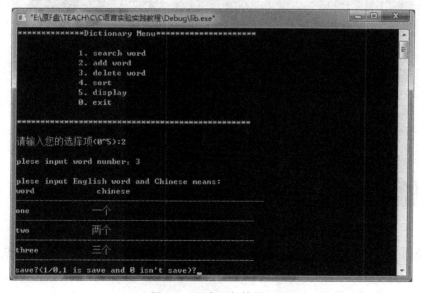

图3.4　程序运行结果1

(3) 上面输入了三个词条,可以选择菜单项 5 display 看看词典中的内容,输入选项 5,界面如图 3.5 所示。除了显示刚才输入的 3 个词条,同时显示主菜单。

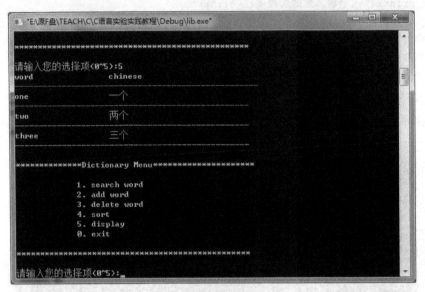

图 3.5 程序运行结果 2

(4) 对刚才输入的 3 个词条按英文单词进行排序,输入选项 4,界面上直接出现已排序好的提示信息,这时可以选择菜单 5 display 看看是否已排序好,排序后再次输入选择项 5,界面如图 3.6 所示。可以看到数据文件已经按照单词排好序了。

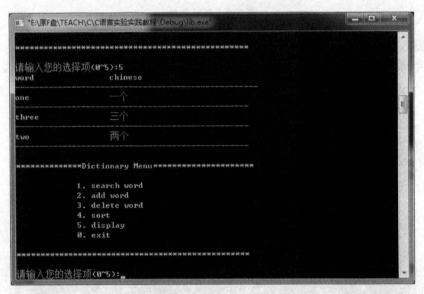

图 3.6 程序运行结果 3

(5) 菜单 1 search word 可以查找单词及其释义,先输入选择项 1,然后输入要查找的单词,这里输入单词 three,结果如图 3.7 所示。除了显示出符合条件的词条外,同时显示出菜单可以继续进行其他的操作。

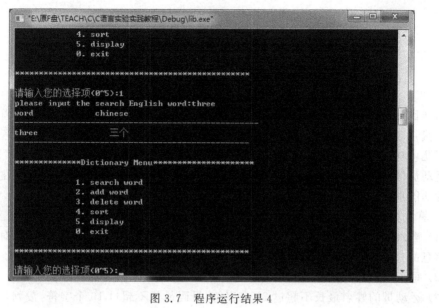

图 3.7　程序运行结果 4

（6）若要结束整个程序的运行，输入选择项 0 即可退出。

项目四　运动会成绩统计与管理

本项目要求实现竞技比赛中运动员成绩的统计与管理。运行程序后，用户首先需要输入各个裁判对运动员的打分，程序会自动将其保存在一个用户指定的文件中。程序要求具有以下几项功能：求出各运动员的总分数、平均分，按姓名、按号码查询其记录并显示，浏览全部运动员的成绩和按总分由高到低显示运动员信息等。用户可以根据主菜单的提示选择不同的功能项。

1．算法分析

（1）输入运动员信息后，需要将其存储在数据库文件中。利用本系统实现对文件中运动员信息的查找、浏览全部运动员的成绩和按总分由高到低显示运动员的信息。

（2）每个运动员的信息包括运动员的姓名、号码和若干裁判的打分。

（3）运动员的姓名最长不超过 20 个字符，号码最长不超过 10 个字符，裁判员的个数可以根据需要定义。

（4）采用菜单工作方式，在一个操作执行后，继续显示主菜单，根据需要再选择菜单项进行其他操作。当选择退出操作时，退出系统，程序运行结束。

根据以上算法分析，该系统的程序主框架功能结构如图 4.1 所示。

图 4.1　主框架功能结构图

程序流程图如图 4.2 所示。

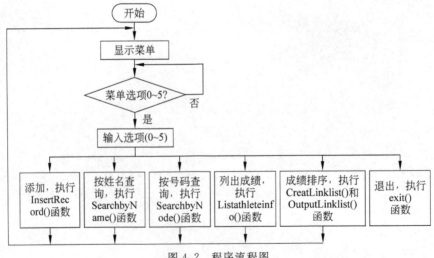

图 4.2　程序流程图

2. 难点提示

(1) 程序要定义一个全局结构体类型 struct AthleteScore,它有 4 个成员：运动员姓名 name、运动员号码 code、裁判打分 score 以及总分 total,其中前三项都为数组,裁判人数可根据具体情况设定 score 数组长度即可。

(2) 定义一个全局结构体类型 struct LinkNode,它有 5 个成员,其中前 4 个成员与 struct AthleteScore 相同,另外增加了一个结构体类型的指针变量 struct LinkNode * next,这个结构体类型用于创建链表数据结构来存放运动员的信息。

(3) 本程序采用模块化的程序设计方法。其中实现的功能模块有：添加模块、按姓名查询模块、按号码查询模块、列出成绩模块和成绩排序模块。添加模块的作用是输入运动员的各项信息到指定的文件中,程序通过函数 InsertRecord 来实现此功能,先以追加的方式打开指定文件,然后通过循环语句录入每个运动员的各项信息,所有运动员的信息一旦录完,调用相关函数写入指定文件；程序实现两种查询方式,一种是按运动员姓名查询,另一种是按号码来查询,分别通过两个函数来实现：SearchbyName(char * fname,char * key)和 SearchbyCode(char * fname, * key),函数 SearchbyName 的作用是按名字来查询运动员的信息,其中参数 key 指定了要查询的运动员的姓名,函数 SearchbyCode 的作用是按号码来查询运动员的信息,其中参数 key 指定了要查询的运动员的号码；列出成绩模块的作用是列出所有运动员的信息,通过函数 OutputLinklist 来实现此项功能,以读的方式打开存放运动员信息的文件,将文件中的内容读入内存,再将内存中的内容输出到屏幕上；成绩排序模块的作用是按总分数由高到低的顺序来对运动员成绩进行排序,并依次输出,通过函数 OutputLinklist 来实现此项功能,先以读的方式打开存放运动员信息的文件,通过 malloc 函数开辟动态存储空间,并形成链表,将文件中的信息按照运动员成绩总分由高到低读取到链表中,再顺序显示链表中各结点的信息。

(4) 采用菜单工作方式,在一个操作执行后,自动显示菜单,可以继续选择另一个操作项；当选择退出操作项时,退出系统,整个程序运行结束。

3. 部分代码

这里只给出主函数和各模块调用函数的代码,各函数分别又调用了其他函数,其他函数请读者自己根据以上提示信息写出。

(1) 主函数代码

```
int main()
{   int c;
    char buf[BUFFSIZE];
    while(1)
    {   printf("\n--Athlete System,please input a select --\n");
        printf("|  1 : insert record to a file.          |\n");
        printf("|  2 : search record by name.            |\n");
        printf("|  3 : search record by code.            |\n");
        printf("|  4 : list all the records.             |\n");
        printf("|  5 : sort the records by total.        |\n");
        printf("|  0 : quit.                             |\n");
        printf("------------------------------\n");
        printf("Please input a command:\n");
```

```
            scanf("%d",&c);                          /*输入选择命令*/
            switch(c)
            { case 1:
                  InsertRecord();
                  getchar();
                  break;
              case 2:                                /*按运动员的姓名寻找记录*/
                  printf("Please input the athlete's name:\n");
                  scanf("%s",buf);
                  SearchbyName(fname,buf);
                  getchar();break;
              case 3:                                /*按运动员的号码寻找记录*/
                  printf("Please input the athlete's code:\n");
                  scanf("%s",buf);
                  SearchbyCode(fname,buf);
                  getchar();break;
              case 4:                                /*列出所有运动员记录*/
                  Listathleteinfo(fname);
                  getchar();break;
              case 5:                                /*按总分从高到低排列显示*/
                  if((head=CreatLinklist(fname))!=NULL)
                      OutputLinklist(head);
                  getchar();break;
              case 0:exit(0);
              default: break;
            }
        }
        return 1;
    }
```

(2) 添加模块函数代码

```
void InsertRecord()
{   FILE * fp;
    char i,j,n;
    struct AthleteScore s;
    fp=fopen(fname,"a+");
    printf("Please input the record number : ");
    scanf("%d",&n);
    for(i=0;i<n;i++)
    {   printf("Input the athlete's name: ");
        scanf("%s",&s.name);
        printf("Input the athlete's code: ");
        scanf("%s",&s.code);
        for(j=0;j<JUDEGNUM;j++)
        {   printf("Input the %s mark: ",judgement[j]);
            scanf("%d",&s.score[j]);
        }
```

```c
        PutRecord(fp,&s);
    }
    fclose(fp);
}
```

(3) 查询模块函数代码

```c
/*按运动员姓名查找记录*/
int SearchbyName(char * fname, char * key)
{   FILE * fp;
    int c;
    struct AthleteScore s;
    if((fp=fopen(fname,"r"))==NULL)
    {   printf("Can't open file %s.\n",fname);
        return 0;
    }
    c=0;
    while(GetRecord(fp,&s)!=0)
    {   if(strcmp(s.name,key)==0)
        {   ShowAthleteRecord(&s);   c++;   }
    }
    fclose(fp);
    if(c==0)
        printf("The athlete %s is not in the file %s.\n",key,fname);
    return 1;
}
/*按运动员号码查找记录*/
int SearchbyCode(char * fname, char * key)
{   FILE * fp;
    int c;
    struct AthleteScore s;
    if((fp=fopen(fname,"r"))==NULL)
    {   printf("Can't open file %s.\n",fname);
        return 0;
    }
    c=0;
    while(GetRecord(fp,&s)!=0)
    {   if(strcmp(s.code,key)==0)
        {   ShowAthleteRecord(&s);
            c++;
            break;
        }
    }
    fclose(fp);
    if(c==0)
        printf("The athlete %s is not in the file %s.\n",key,fname);
```

```
        return 1;
    }
```

(4) 列出成绩模块函数代码

```
/*列表显示运动员成绩*/
void Listathleteinfo(char * fname)
{   FILE * fp;
    struct AthleteScore s;
    if((fp=fopen(fname,"r"))==NULL)
    {   printf("Can't open file %s.\n",fname);return;   }
    while(GetRecord(fp,&s)!=0)
        ShowAthleteRecord(&s);
    fclose(fp);
    return;
}
```

(5) 成绩排序模块代码

```
/*顺序显示链表各表元*/
void OutputLinklist(struct LinkNode * h)
{   while(h!=NULL)
    {   ShowAthleteRecord((struct AthleteScore *)h);
        printf("\n");
        while(getchar()!='\n');
        h=h->next;
    }
    return;
}
```

4. 运行结果示例

(1) 程序运行时,显示菜单如图 4.3 所示。

图 4.3 显示菜单

(2) 输入菜单项命令 1 后，系统会问输入几个运动员成绩，这里输入 3 个，开始输入每个运动员的各项信息：姓名、号码和各裁判给的成绩，3 个运动员的信息输入结束后，系统会自动将这些信息存入指定的文件中，接着又显示菜单，可以进行其他操作，界面如图 4.4 所示。

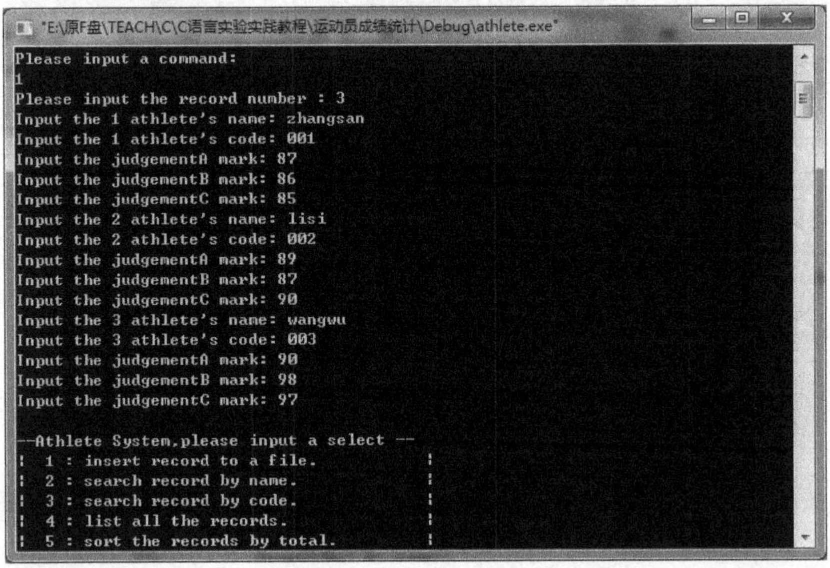

图 4.4 运行界面

(3) 再输入菜单项命令 2，按运动员姓名查询，接下来系统要求输入运动员姓名，这里输入 lisi，系统从文件中调出符合查询条件的信息，同时显示菜单，可以进行其他操作，显示界面如图 4.5 所示。

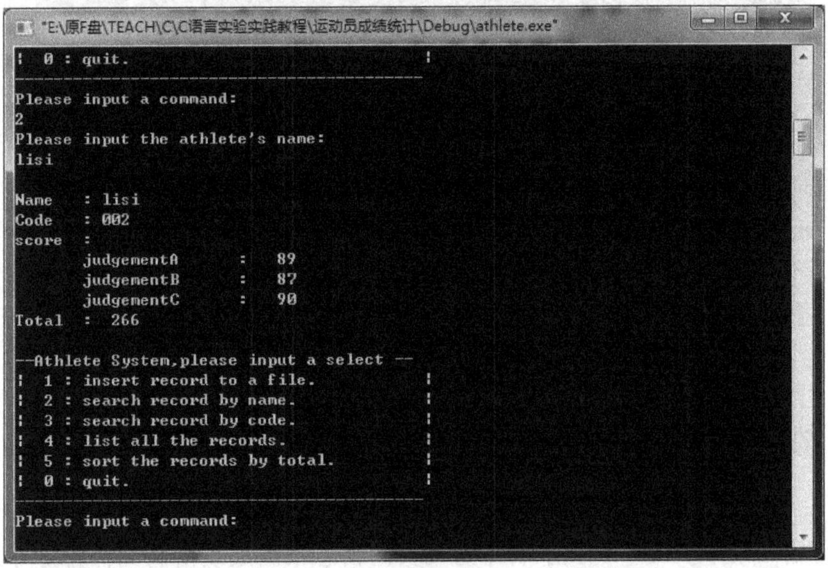

图 4.5 显示界面

（4）再输入菜单项命令3，按运动员号码查询，接下来系统要求输入运动员号码，这里输入003，系统从文件中调出符合查询条件的信息，同时显示菜单，可以进行其他操作，显示界面如图4.6所示。

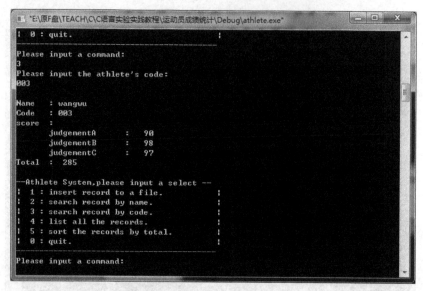

图4.6　显示界面

（5）再输入菜单项命令4，列出所有运动员的信息，同时显示菜单，可以进行其他操作，界面如图4.7所示。

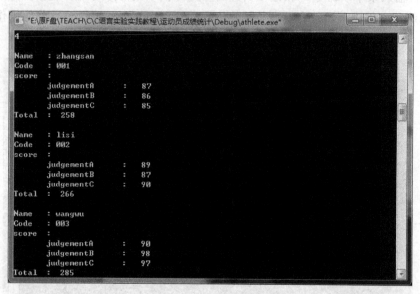

图4.7　运行界面

（6）再输入菜单项命令5，对文件中运动员的信息根据成绩由高到低排好序后，输出到屏幕上，同时显示菜单，可以继续进行其他操作。输入菜单项命令0，退出系统，程序运行结束。

项目五 俄罗斯方块

俄罗斯方块(Tetris,俄文:Tетрис)是20世纪80年末期至20世纪90年代初期风靡全球的电子游戏,是落下型益智游戏的始祖,是一款风靡全球的电视游戏机和掌上游戏机游戏,它曾经造成的轰动与造成的经济价值可以说是游戏史上的一件大事。这款游戏最初是由前苏联的游戏制作人Alex Pajitnov制作的,它的规则简单,容易上手,且游戏过程变化无穷,令人上瘾,现在要求实现一款在电脑上运行的俄罗斯方块游戏。

1. 算法分析

本游戏要求实现以下基本功能和算法:

(1) 控制方块的左右运动,通过键盘左右方向箭头("◀"和"▶")来实现。

(2) 控制方块向下的加速运动,可以使方块快速下落到底部,通过向下方向箭来实现。

(3) 控制方块的旋转变换,通过空格键来实现。

(4) 退出游戏,通过 Esc 键来实现。

2. 难点提示

(1) 设计这个游戏的一个关键点是如何表示方块,方块有7种基本形状,所有的形状都可以放在4×4的格子里,如图5.1所示,程序中可以使用一个三维数组BOX来表示图5.1所示的7种基本的方块形状。这7种形状又可以进行旋转得到不同的形状。假定旋转的方向是逆时针方向(顺时针方向道理一样),形状(a)可旋转为如图5.2所示(a1)、(a2)、(a3)三种形状,形状(b)可旋转为如图5.3所示(b1)一种形状,形状(c)可旋转为如图5.4所示(c1)一种形状,形状(d)旋转后不产生新的图形,形状(e)可旋转为如图5.5所示(e1)一种形状,形状(f)可旋转为如图5.6所示(f1)、(f2)、(f3)三种形状,形状(g)可旋转为如图5.7所示(g1)、(g2)、(g3)三种形状。这7种形状及它们旋转后的变形体总共有19种形状。

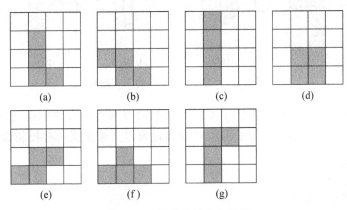

图5.1 7种基本的方块形状

```
Int BOX[7][4][4]={{{1,1,1,1},{0,0,0,0},{0,0,0,0},{0,0,0,0}},
                  {{1,1,1,0},{1,0,0,0},{0,0,0,0},{0,0,0,0}},
                  {{1,1,1,0},{0,0,1,0},{0,0,0,0},{0,0,0,0}},
```

```
        {{1,1,1,0},{0,1,0,0},{0,0,0,0},{0,0,0,0}},
        {{1,1,0,0},{0,1,1,0},{0,0,0,0},{0,0,0,0}},
        {{0,1,1,0},{1,1,0,0},{0,0,0,0},{0,0,0,0}},
        {{1,1,0,0},{1,1,0,0},{0,0,0,0},{0,0,0,0}}
        }
```

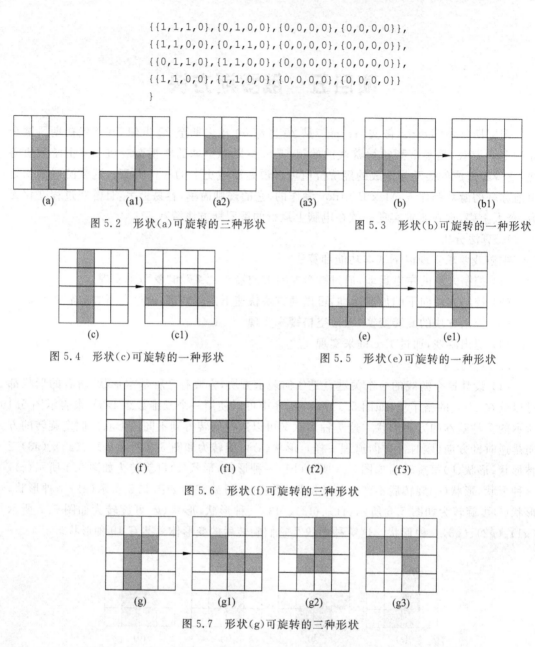

图5.2 形状(a)可旋转的三种形状   图5.3 形状(b)可旋转的一种形状

图5.4 形状(c)可旋转的一种形状   图5.5 形状(e)可旋转的一种形状

图5.6 形状(f)可旋转的三种形状

图5.7 形状(g)可旋转的三种形状

(2) 在运行游戏时,用户需要使用不同的控制命令来控制方块的运动,如左右运动、方块的翻转变形等。为了控制方块的运动,可以通过宏定义来定义一些控制命令。

```
/*重画界面命令*/
#define CMDDRAW      5
/*消去一个满行的命令*/
#define CMDDELLINE   6
/*自动下移一行的命令*/
#define CMDAOTODOWN  7
/*生产新的方块*/
#define CMDGEN       8
```

```c
/*向左移动的命令,以左箭头<-控制,它的ASCII码值是75*/
#define CMDLEFTMOVE     75
/*向右移动的命令,以右箭头->控制,它的ASCII码值是77*/
#define CMDRINGHTMOVE   77
/*旋转方块的命令,以空格来控制*/
#define CMDROTATE       57
/*向下移动的命令,以向下的箭头控制,它的ASCII码值是80*/
#define CMDDOWNMOVE     80
/*退出游戏的控制命令,以Esc键控制,它的ASCII码值是1*/
#define CMDESC          1
```

(3) 为了实现以上这些命令,分别编写相应的函数来实现。

3. 程序代码

这里只给出主函数和一部分函数的代码,框架中的其他函数请读者参照所给部分代码写出。

```c
/*定义了方块下降的时间间隔*/
#define TIMEINTERVAL    4
/*下面定义了游戏区的大小*/
#define MAXWIDTH        15
#define MAXHEIGHT       30
/*组成方块的小格子的宽度*/
#define BOXWIDTH        15
/*用两个数组来表示新旧两个矩形游戏区*/
int oldarea[MAXHEIGHT+1][MAXWIDTH];
int area[MAXHEIGHT+1][MAXWIDTH];
/*定义一个需要改变的屏幕区域,除此之外的区域不用进行重绘*/
int actW,actH,actX,actY;
/*当前方块的一些属性(坐标,颜色,高,宽)*/
int curX,curY,curColor,curW,curH;
/*新的方块的一些属性(坐标,颜色,高,宽)*/
int newX,newY,newColor,newW,newH;
/*制定方块的状态*/
int active;
/*存储当前方块的数组*/
int box[4][4];
/*当前方块的颜色*/
int BOXCOLOR;
/*控制命令*/
int CMD;
/*定义7种基本的方块形状*/
int BOX[7][4][4]={
{{1,1,1,1},{0,0,0,0},{0,0,0,0},{0,0,0,0}},
{{1,1,1,0},{1,0,0,0},{0,0,0,0},{0,0,0,0}},
{{1,1,1,0},{0,0,1,0},{0,0,0,0},{0,0,0,0}},
{{1,1,1,0},{0,1,0,0},{0,0,0,0},{0,0,0,0}},
{{1,1,0,0},{0,1,1,0},{0,0,0,0},{0,0,0,0}},
{{0,1,1,0},{1,1,0,0},{0,0,0,0},{0,0,0,0}},
```

```c
  {{1,1,0,0},{1,1,0,0},{0,0,0,0},{0,0,0,0}}
};
/*得到方块的宽度,即从右向左第一个不空的列*/
int GetWidth()
{   int i,j;
    for(i=3;i>0;i--)
      for(j=0;j<4;j++)
        if(box[j][i]) return i;
    return 0;
}
/*得到方块的高度,从上往下第一个不空的行*/
int GetHeight()
{   int i,j;
    for(j=3;j>0;j--)
      for(i=0;i<4;i++)
        if(box[j][i]) return j;
    return 0;
}
/*清除原有的方块占有的空间*/
void ClearOldspace()
{   int i,j;
    for(j=0;j<=curH;j++)
      for(i=0;i<=curW; i++)
          if(box[j][i])area[curY+j][curX+i]=0;
}
/*置位新方块的位置*/
void PutNewspace()
{   int i,j;
    for(j=0;j<=newH;j++)
      for(i=0;i<=newW;i++)
        if(box[j][i])
          area[newY+j][newX+i]=BOXCOLOR;
}
/*判断方块的移动是否造成区域冲突*/
int MoveCollision(int box[][4])
{   int i,j;
    if(newX<0) return 1;
    if(newX+newW>=MAXWIDTH) return 1;
    if(newY<0) return 1;
    for(j=0;j<=newH;j++)
      for(i=0;i<=newW;i++)
         if(area[newY+j][newX+i]&&box[j][i]) return 1;
    return 0;
}
/*判断翻转方块是否造成区域的冲突*/
int RotateBoxCollision(int box[][4])
{   int i,j;
```

```
    if(newX+newW>=MAXWIDTH) newX=MAXWIDTH-1-newW;
    if(newY+newH>=MAXHEIGHT) newY=MAXHEIGHT-1-newH;
    if(MoveCollision(box)) return 1;
    for(i=0;i<=newW;i++)
    for(j=0;j<=newH;j++)
    if(area[newY+j][newX+i])
       newX-=newW-i+1; goto L;
    L: return MoveCollision(box);
}
/*游戏结束*/
int GameOver()
{   if(!active &&(curY+curH>MAXHEIGHT-3)) return 1;
    else   return 0;
}
/*判断是否超时,即是否超过允许的时间间隔*/
int TimeOut()
{   static long tm,old;
    tm=biostime(0,tm);
    if(tm-old<TIMEINTERVAL) return 0;
    else { old=tm;    return 1;   }
}
/*重绘游戏区*/
void DrawSpace()
{   int row,col,x1,y1,x2,y2;
    for(row=actY;row<=actY+actH;row++)
    for(col=actX;col<=actX+actW;col++)
       if(area[row][col]!=oldarea[row][col])
       {   if(area[row][col]==0)
             setfillstyle(SOLID_FILL,BLACK);
           else  setfillstyle(SOLID_FILL,BOXCOLOR);
           x1=56+col*BOXWIDTH;
           x2=x1+BOXWIDTH;
           y1=464-(row+1)*BOXWIDTH;
           y2=y1+BOXWIDTH;
           bar(++x1,++y1,--x2,--y2);
           oldarea[row][col]=area[row][col];
       }
    CMD=0;
}
/*消去满行*/
void ClearFullline()
{   int row,col, rowEnd,full,i,j;
    rowEnd=newY+newH;
    if(rowEnd>=MAXHEIGHT-1) rowEnd=MAXHEIGHT-2;
    for(row=newY; row<=rowEnd;)
    {   full=1;
        for(col=0;col<MAXWIDTH;col++)
```

```c
            if(!area[row][col]){ full=0; break; }
        if(!full)
        {  ++row;  continue;  }
        for(j=row;j<MAXHEIGHT-1;j++)
          for(i=0;i<MAXWIDTH;i++)
            area[j][i]=area[j+1][i];
        actX=0;
        actY=row;
        actW=MAXWIDTH-1;
        actH=MAXHEIGHT-1-row;
        DrawSpace();
        rowEnd--;
    }
    CMD=CMDGEN;
}
/*向左移动方块*/
int MoveLeft()
{   newX=curX-1;
    ClearOldspace();
    if(MoveCollision(box))
    {   newX=curX;
        PutNewspace();
        CMD=0;
        return 0;
    }
    PutNewspace();
    actW=curW+1;
    actX=curX=newX;
    CMD=CMDDRAW;
    return 1;
}
/*向右移动方块*/
int MoveRight()
{   newX=curX+1; ClearOldspace();
    if(MoveCollision(box))
    {   newX=curX;
        PutNewspace();
        CMD=0;
        return 0;
    }
    PutNewspace();
    actW=curW+1;
    actX=curX;
    curX=newX;
    CMD=CMDDRAW;
```

```c
        return 1;
}
/*向下移动方块*/
int MoveDown()
{   int i,j;
    newY=curY-1;
    ClearOldspace();
    if(MoveCollision(box))
    {   newY=curY;
        PutNewspace();
        active=0;
        CMD=CMDDELLINE;
        return 0;
    }
    PutNewspace();
    actH=curH+1;
    actY=newY;
    curY=newY;
    CMD=CMDDRAW;
    return 1;
}
/*按加速键后方块迅速下落到底*/
void MoveBottom()
{   while(active)
    {   MoveDown();
        DrawSpace();
    }
    CMD=CMDDELLINE;
}
/*初始化*/
void InitialGame()
{   int i,j,x1,y1,x2,y2;
    int driver=DETECT, mode=0;
    initgraph(&driver,&mode,"e:\\tc");
    cleardevice();
    randomize();
    setfillstyle(SOLID_FILL,BLUE);
    bar(0,0,639,479);
    x1=56;
    y1=464-BOXWIDTH*MAXHEIGHT;
    x2=56+MAXWIDTH*BOXWIDTH;
    y2=464;
    rectangle(--x1,--y1,++x2,++y2);
    setfillstyle(SOLID_FILL,BLACK);
    bar(++x1,++y1,--x2,--y2);
```

```c
        y1=464-MAXHEIGHT*BOXWIDTH; y2=464;
        setcolor(DARKGRAY);
        for(i=0;i<MAXWIDTH;i++)
        {   x1=56+i*BOXWIDTH;
            line(x1,y1,x1,y2);
        }
        x1=56; x2=x1+MAXWIDTH*BOXWIDTH;
        for(j=0;j<MAXHEIGHT;j++)
        {   y1=464-j*BOXWIDTH;
            line(x1,y1,x2,y1);
        }
        for(j=0;j<MAXHEIGHT;j++)
          for(i=0;i<MAXWIDTH;i++)
             area[j][i]=oldarea[j][i]=0;
        actX=0; actY=0; actW=MAXWIDTH-1; actH=MAXHEIGHT-1;
        DrawSpace();
        CMD=CMDGEN;
}
/*得到控制命令*/
void GetCMD()
{   if(CMD) return;
    if(TimeOut())
    {   CMD=CMDAOTODOWN;
        return;
    }
    if(bioskey(1))
    {   CMD=bioskey(0)>>8;
        return;
    }
}
/*生成一个新的方块*/
int GenerateNewbox()
{   int i,j,boxidx;
    boxidx=random(7); BOXCOLOR=random(7)+1;
    for(j=0;j<4;j++)
      for(i=0;i<4;i++)
        box[j][i]=BOX[boxidx][j][i];
    curW=GetWidth(); curH=GetHeight();
    curX=(MAXWIDTH+curW)/2;
    if(curX+curW>=MAXWIDTH)curX=MAXWIDTH-1-curW;
    curY=MAXHEIGHT-1-curH;
    newX=curX; newY=curY; actX=curX;actY=curY;
    actW=newW=curW; actH=newH=curH;
    active=1;
    if(MoveCollision(box)) return 0;
```

```c
        PutNewspace();
        DrawSpace(); CMD=0;
        return 1;
    }
/* 翻转方块 */
int RotateBox()
{   int newBox[4][4];
    int i,j;
    ClearOldspace();
    for(j=0;j<4;j++)
    for(i=0;i<4;i++)
    newBox[j][i]=0;
    for(j=0;j<4;j++)
      for(i=0;i<4;i++)
          newBox[curW-i][j]=box[j][i];
    newW=curH;
    newH=curW;
    if(RotateBoxCollision(newBox))
        {   newW=curW;
            newH=curH;
            newX=curX;
            newY=curY;
            PutNewspace();
            CMD=0;
            return 0;
        }
    for(j=0;j<4;j++)
      for(i=0;i<4;i++)
          box[j][i]=newBox[j][i];
    PutNewspace();
    actH=newH>curH?newH:curH;
    actW=curX+actH-newX;
    actX=newX;
    actY=newY;
    curX=newX;
    curY=newY;
    curW=newW;
    curH=newH;
    CMD=CMDDRAW;
    return 1;
}
/* 根据获得的命令来执行不同的操作 */
void ExecuteCMD()
{   switch(CMD)
        {   case CMDLEFTMOVE: MoveLeft();        break;
```

```
            case CMDRINGHTMOVE: MoveRight();    break;
            case CMDAOTODOWN: MoveDown();       break;
            case CMDROTATE: RotateBox();        break;
            case CMDDOWNMOVE: MoveBottom();     break;
            case CMDDRAW: DrawSpace();          break;
            case CMDDELLINE: ClearFullline();   break;
            case CMDGEN: GenerateNewbox();      break;
            case CMDESC: closegraph();return 0;
            default: CMD=0;
        }
    }
}
/*主函数*/
void main()
{   int i;
    InitialGame();
    do
    {   GetCMD();  ExecuteCMD(); }
    while(!GameOver());
    getch();
    closegraph();
    return 0;
}
```

4. 运行结果示例

（1）游戏开始运行时，从顶部下来第一个方块，方块的形状和颜色随机产生，每次下来的可能不一样，也可能一样，界面如图5.8所示。

（2）上面方块下落到底部后，顶部会随机出现第二个方块，如图5.9所示。

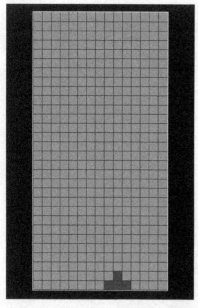
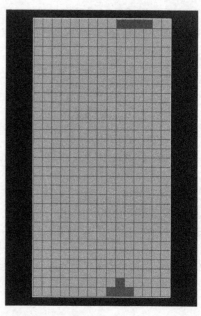

图5.8 游戏运行后第一个方块下落至底部　　　图5.9 游戏运行中出现第二个方块

(3) 当底部一行格子全部充满时会自动消去这行,同时顶部有新的方块下落,如图 5.10 所示。

(4) 当某一列的方格堆积到顶部时,游戏结束,如图 5.11 所示。

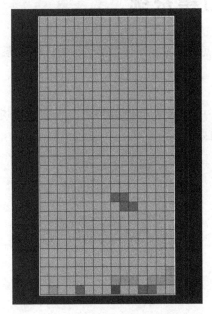

图 5.10　游戏运行中底部满行消除后　　　图 5.11　游戏运行结束

项目六　火车订票系统

本系统要求实现的功能有：添加火车的车次信息，查询火车车次信息，预订火车票，更新火车车次信息，保存信息到文件等。

1. 算法分析

（1）为了方便管理火车的车次信息，可以在程序中定义火车的结构体 train。另外定义结构体 man 来表示订票人的信息。它们的定义分别如下：

```
/*定义存储火车信息的结构体*/
struct train
{   char num[10];                    /*列车号*/
    char city[10];                   /*目的城市*/
    char takeoffTime[10];            /*发车时间*/
    char receiveTime[10];            /*到达时间*/
    int  price;                      /*票价*/
    int  bookNum ;                   /*票数*/
};
/*订票人的信息*/
struct man
{   char num[10];                    /*ID*/
    char name[10];                   /*姓名*/
    int  bookNum ;                   /*需求的票数*/
};
```

（2）使用结构体存储每辆火车的信息，并使用链表来存储全部火车信息，在保存时将链表中的火车信息保存到文件数据库 train.txt 中。定义火车信息链表的结点结构如下：

```
struct node
{   struct train data ;
    struct node * next ;
}
```

（3）使用结构体存储每个订票人的信息，并使用链表来存储全部订票人的信息，在保存时将链表中的订票人信息保存到文件数据库 man.txt 中。定义订票人链表的结点结构如下：

```
struct people
{   struct man data ;
    struct people * next ;
}
```

（4）将整个系统分为以下 6 个模块来实现：添加车次信息模块、查询车次信息模块、订票模块、车次信息更新模块、系统推荐车次模块以及信息保存模块。

根据以上算法分析,该系统的程序主框架功能结构如图6.1所示。

图6.1 程序主框架功能结构图

程序流程图如图6.2所示。

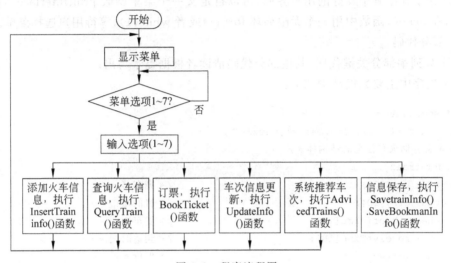

图6.2 程序流程图

2. 难点提示

(1) 本系统对于火车车次信息的管理和订票用户信息的管理都采用链表来实现,为了从文件中读取火车信息和订票人信息,需要定义这两个信息的全局变量,其中包括指向火车信息的指针变量和指向订票人信息的指针变量,从而能够方便地读取这些信息到链表结构中,各功能模块的实现主要采用了链表的相关操作。

(2) 程序除main()函数外,另外有6个函数分别用来实现以上6个模块功能。添加车次信息模块实现添加火车车次信息的功能,这些信息包括火车的车次、目的地、发车时间、到达时间、票价和票数等,定义函数 void InsertTraininfo(Link linkhead)来实现此模块的功能,接受的参数是一个链表的头指针linkhead。车次信息查询模块的功能是查询车次的信息,可以提供多种查询方式,如按车次查询和按到达的目的地查询,定义函数 void QueryTrain(Link linkhead)来实现此模块的功能,它接受的参数也是一个链表的头指针linkhead。订票模块的功能是为用户订购指定车次的票,需要用户输入要到达的城市,系统会自动将相关车次及其信息列出,用户再选择一个车次进行预订,此时需要输入用户的姓名、ID以及订票的张数等信息,定义函数 void BookTicket(Link l, bookManLink k)来实现此模块的功能,接受的参数是两个链表的头指针l和k,前者是车次信息链表的头指针,后者是订票人信息链表的头指针。车次信息更新模块的功能是更新需要变更的车次信息,用户需要首先给出原来的车次信息,然后再输入更改后的信息,系统将自动更新这些信息,定义函数 void UpdateInfo(Link linkhead)来实现此模块的功能,接受的参数 linkhead 也是一个链表的头指针。系统推荐车次模块的作用是系统自动为用户推荐可乘的车次,用户需要输入要到达的城市,系统会自动给用户寻找可乘的车次,并提示给用户,定义函数 void AdvicedTrains(Link linkhead)来实现此模块的功能,接受的参数 linkhead 也是一个链表的头指针。信息保存模块的作用是将用户此次进入系统时的相关操作进行保存,使用两个文件"train.txt"和"man.txt"来分别保存火车的车次信息和订票的用户信息,定义两个函数 void SaveTrainInfo(Link l)和 void SaveBookmanInfo(bookManLink k)来实现此模块的功能。

(3) 为了程序有个友好的用户界面,可以自定义一个菜单函数 printInterface()负责显示菜单,在 main()函数中用一个无限循环 for(;;)或者 while(1),等待用户选择菜单。

3. 部分代码

以下只列举部分关键代码,其他部分代码请读者根据提示写出。

(1) 程序中主要数据结构如下:

```
int shoudsave=0 ;
int count1=0,count2=0,mark=0,mark1=0 ;
/*定义存储火车信息的结构体*/
struct train
{    char num[10];                      /*列车号*/
     char city[10];                     /*目的城市*/
     char takeoffTime[10];              /*发车时间*/
     char receiveTime[10];              /*到达时间*/
     int  price;                        /*票价*/
     int  bookNum ;                     /*票数*/
};
/*订票人的信息*/
struct man
{    char num[10];                      /*ID*/
     char name[10];                     /*姓名*/
     int  bookNum;                      /*需求的票数*/
};
/*定义火车信息链表的结点结构*/
typedef struct node
{    struct train data;
     struct node * next;
}Node, * Link;
/*定义订票人链表的结点结构*/
typedef struct people
{    struct man data;
     struct people * next;
}bookMan, * bookManLink;
```

(2) 添加车次信息模块代码

```
void InsertTraininfo(Link linkhead)
```

```c
{   struct node *p,*r,*s ;
    char num[10];
    r=linkhead ;
    s=linkhead->next ;
    while(r->next!=NULL)
    r=r->next ;
    while(1)
    {   printf("please input the number of the train(0-return)");
        scanf("%s",num);
        if(strcmp(num,"0")==0)break ;
        /*判断是否已经存在*/
        while(s)
        {   if(strcmp(s->data.num,num)==0)
            {   printf("the train '%s'has been born!\n",num);
                return;
            }
            s=s->next ;
        }
        p=(struct node *)malloc(sizeof(struct node));
        strcpy(p->data.num,num);
        printf("Input the city where the train will reach:");
        scanf("%s",p->data.city);
        printf("Input the time which the train take off:");
        scanf("%s",p->data.takeoffTime);
        printf("Input the time which the train receive:");
        scanf("%s",&p->data.receiveTime);
        printf("Input the price of ticket:");
        scanf("%d",&p->data.price);
        printf("Input the number of booked tickets:");
        scanf("%d",&p->data.bookNum);
        p->next=NULL ;
        r->next=p ;
        r=p ;
        shoudsave=1 ;
    }
}
```

(3) 车次信息查询模块代码

```c
/*查询火车信息*/
void QueryTrain(Link l)
{   Node *p ;
    int sel ;
    char str1[5],str2[10];
    if(!l->next)
    {   printf("There is not any record !");
```

```
            return ;
        }
        printf("Choose the way:\n>>1:according to the number of train;\n>>2:according to the city:\n");
        scanf("%d",&sel);
        if(sel==1)
        {   printf("Input the the number of train:");
            scanf("%s",str1);
            p=Locate1(l,str1,"num");
            if(p)printTrainInfo(p);
            else
            {   mark1=1 ;
                printf("\nthe file can't be found!");
            }
        }
        else if(sel==2)
        {   printf("Input the city:");
            scanf("%s",str2);
            p=Locate1(l,str2,"city");
            if(p)  printTrainInfo(p);
            else
            {   mark1=1 ;
                printf("\nthe file can't be found!");
            }
        }
    }
```

(4) 订票模块代码

```
/*订票子模块*/
void BookTicket(Link l,bookManLink k)
{   Node * r[10], * p ;
    char ch,dem ;
    bookMan * v, * h ;
    int i=0,t=0 ;
    char str[10],str1[10],str2[10];
    v=k ;
    while(v->next!=NULL)
    v=v->next ;
    printf("Input the city you want to go: ");
    scanf("%s",&str);
    p=l->next ;
    while(p!=NULL)
    {   if(strcmp(p->data.city,str)==0)
        {   r[i]=p;    i++;    }
        p=p->next ;
```

```
        }
    printf("\n\nthe number of record have %d\n",i);
    for(t=0;t<i;t++)
        printTrainInfo(r[t]);
    if(i==0)
        printf("\n\t\tSorry!Can't find the train for you!\n");
    else
    {   printf("\ndo you want to book it?<1/0>\n");
        scanf("%d",&ch);
        if(ch==1)
        {   h=(bookMan*)malloc(sizeof(bookMan));
            printf("Input your name: ");
            scanf("%s",&str1);
            strcpy(h->data.name,str1);
            printf("Input your id: ");
            scanf("%s",&str2);
            strcpy(h->data.num,str2);
            printf("Input your bookNum: ");
            scanf("%d",&dem);
            h->data.bookNum=dem ;
            h->next=NULL ;
            v->next=h ;
            v=h ;
            printf("\nLucky!you have booked a ticket!");
            getch();
            shoudsave=1 ;
        }
    }
}
```

(5) 车次信息更新模块代码

```
/*修改火车信息*/
void UpdateInfo(Link l)
{   Node * p ;
    char findmess[20],ch ;
    if(!l->next)
    {   printf("\nthere isn't record for you to modify!\n");
        return ;
    }
    else
    {   QueryTrain(l);
        if(mark1==0)
        {   printf("\nDo you want to modify it? \n");
            getchar();
            scanf("%c",&ch);
```

```
            if(ch=='y');
        {   printf("\nInput the number of the train:");
            scanf("%s",findmess);
            p=Locate1(l,findmess,"num");
            if(p)
            {   printf("Input new number of train:");
                scanf("%s",&p->data.num);
                printf("Input new city the train will reach:");
                scanf("%s",&p->data.city);
                printf("Input new time the train take off");
                scanf("%s",&p->data.takeoffTime);
                printf("Input new time the train reach:");
                scanf("%s",&p->data.receiveTime);
                printf("Input new price of the ticket::");
                scanf("%d",&p->data.price);
                printf("Input new number of people who have booked ticket:");
                scanf("%d",&p->data.bookNum);
                printf("\nmodifying record is sucessful!\n");
                shoudsave=1 ;
            }
            else
                printf("\t\t\tcan't find the record!");
        }
    }
    else   mark1=0 ;
    }
}
```

(6) 系统推荐车次模块代码

```
/* 系统给用户的提示信息 */
void AdvicedTrains(Link l)
{   Node * r ;
    char str[10];
    int mar=0 ;
    r=l->next ;
    printf("Iuput the city you want to go: ");
    scanf("%s",str);
    while(r)
    {   if(strcmp(r->data.city,str)==0&&r->data.bookNum<200)
        {   mar=1 ;
            printf("\nyou can select the following train!\n");
            printf("\n\nplease select the fourth operation to book the ticket!\n");
            printTrainInfo(r);
        }
        r=r->next ;
```

```
    }
    if(mar==0)
    printf("\n\t\t\tyou can't book any ticket now!\n");
```

(7) 信息保存模块代码

```c
/*保存火车信息*/
void SaveTrainInfo(Link l)
{   FILE * fp ;
    Node * p ;
    int count=0,flag=1 ;
    fp=fopen("c:\\train.txt","wb");
    if(fp==NULL)
    {   printf("the file can't be opened!");
        return ;
    }
    p=l->next ;
    while(p)
    {   if(fwrite(p,sizeof(Node),1,fp)==1)
        {   p=p->next;  count++;    }
        else
        {   flag=0;  break;   }
    }
    if(flag)
    {   printf("the number of the record which have been saved is %d\n",count);
        shoudsave=0 ;
    }
    fclose(fp);
}
/*保存订票人的信息*/
void SaveBookmanInfo(bookManLink k)
{   FILE * fp ;
    bookMan * p ;
    int count=0,flag=1 ;
    fp=fopen("c:\\man.txt","wb");
    if(fp==NULL)
    {   printf("the file can't be opened!");
        return ;
    }
    p=k->next ;
    while(p)
    {   if(fwrite(p,sizeof(bookMan),1,fp)==1)
        {   p=p->next;  count++;   }
        else
```

```
            { flag=0;   break;   }
        }
        if(flag)
        { printf("the number of the record which have been saved is %d\n",count);
            shoudsave=0 ;
        }
        fclose(fp);
}
```

(8) 主函数代码

```
void main()
{   FILE * fp1, * fp2 ;
    Node * p, * r ;
    char ch1,ch2 ;
    Link l ;
    bookManLink k ;
    bookMan * t, * h ;
    int sel ;
    l=(Node * )malloc(sizeof(Node));
    l->next=NULL ;
    r=l ;
    k=(bookMan * )malloc(sizeof(bookMan));
    k->next=NULL ;
    h=k ;
    fp1=fopen("c:\\train.txt","ab+");
    if((fp1==NULL))
    {   printf("can't open the file!");return 0;   }
    while(!feof(fp1))
    {   p=(Node * )malloc(sizeof(Node));
        if(fread(p,sizeof(Node),1,fp1)==1)
        {   p->next=NULL ;
            r->next=p ;
            r=p ;
            count1++;
        }
    }
    fclose(fp1);
    fp2=fopen("c:\\man.txt","ab+");
    if((fp2==NULL))
    {   printf("can't open the file!");return 0;   }
    while(!feof(fp2))
    {   t=(bookMan * )malloc(sizeof(bookMan));
        if(fread(t,sizeof(bookMan),1,fp2)==1)
        {   t->next=NULL ;
            h->next=t ;
            h=t ;
```

```
            count2++;
        }
    }
    fclose(fp2);
    while(1)
    {   clrscr();
        printInterface();
        do{ printf("please choose the operation:  ");
            scanf("%d",&sel);
            clrscr();
        }while(sel<1||sel>7);
        if(sel==8)
        {   if(shoudsave==1)
            {   getchar();
                printf("\nthe file have been changed!do you want to save it(y/n)?\n");
                scanf("%c",&ch1);
                if(ch1=='y'||ch1=='Y')
                {   SaveBookmanInfo(k);
                    SaveTrainInfo(l);
                }
            }
            printf("\nThank you!!You are welcome too\n");
            break ;
        }
        switch(sel)
        {   case 1:InsertTraininfo(l);break ;
            case 2:QueryTrain(l);break ;
            case 3:BookTicket(l,k);break ;
            case 4:UpdateInfo(l);break ;
            case 5: AdvicedTrains(l);break ;
            case 6: SaveTrainInfo(l);SaveBookmanInfo(k);break;
            case 7: return 0;
        }
        printf("\nplease press any key to continue.......");
        getch();
    }
    return 0;
}
```

4. 运行结果示例

(1) 程序运行时,显示功能项菜单如图6.3所示。

(2) 输入选项1,录入火车的信息,可以循环录入多个火车的信息。当停止录入火车信息时,输入0即可结束,如图6.4所示。

(3) 按任意键继续显示功能项菜单,输入选项6可将刚才录入的火车信息存储到磁盘上的文件中,如图6.5所示。上面只录入了两辆火车的信息,没有订票人订票,所示显示保存的火车信息记录数为2,订票人信息记录数为0。

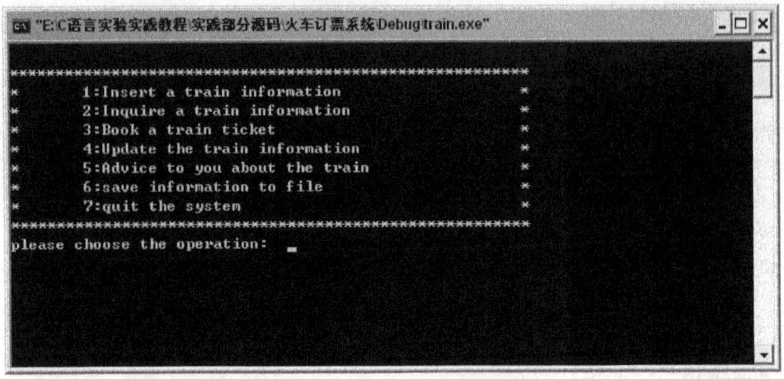

图 6.3 显示功能项菜单

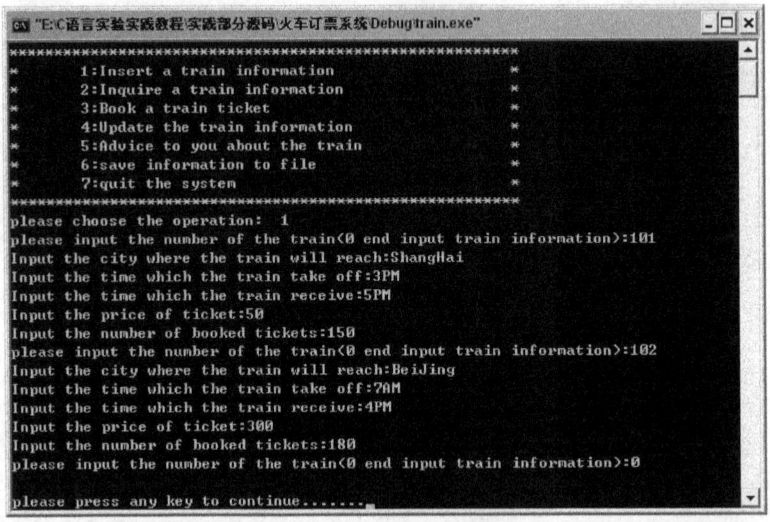

图 6.4 录入火车信息结束时

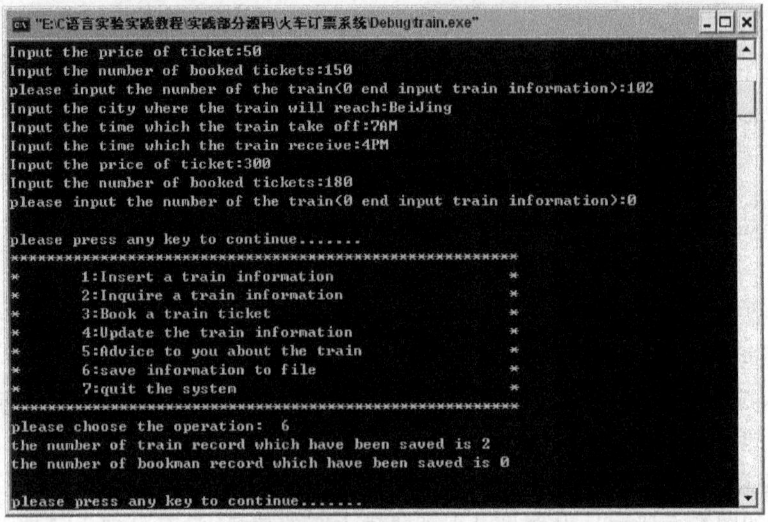

图 6.5 火车信息存储到磁盘上文件后

(4) 按任意键继续显示功能项菜单,此时输入选项 2,可以查询火车信息。可以分别按照火车的车次和到达的城市两个条件进行查询,选 1 按火车车次进行查询,选 2 按火车到达城市进行查询,如图 6.6 所示。

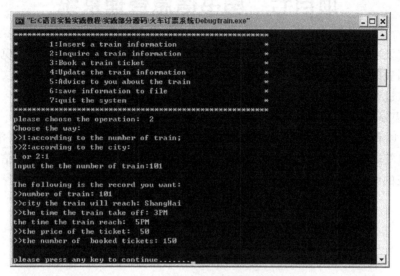

图 6.6 查询火车信息

(5) 按任意键继续显示功能项菜单,此时输入选项 3 可以进行订票操作,按提示输入相关订票信息即可,如图 6.7 所示。

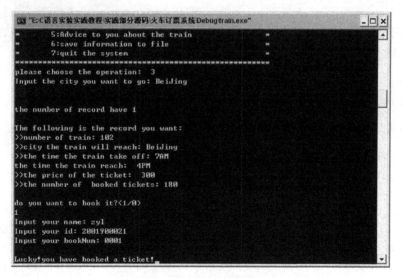

图 6.7 进行订票操作

(6) 按任意键继续显示功能项菜单,此时输入选项 4 可以进行火车信息的更新操作,更新前可以先按照火车车次或者到达城市查询出要更新的记录,然后进行修改。修改结束后选择功能项菜单中的第 6 项将修改后的信息保存即可。

(7) 选择功能菜单中的第 5 项,系统会根据当前火车信息建议订哪个车次的火车票。选择功能菜单中的第 7 项退出系统操作。

项目七　图书信息管理系统

本系统要求设计一个图书信息管理系统,能够完成对图书信息的添加、浏览、查询、删除和对图书列表的清空等操作,可以将图书信息保存在文件数据库中,并可以对文件数据库中的图书信息进行读取与显示。

1. 算法分析

(1) 可以利用结构体来存储每条图书信息,其成员包括图书的书号、书名、作者、出版社和价格信息等,并使用动态链表来保存图书信息,所以需要设计一个存储图书信息的结构体来作为链表的结点,该结构体数据类型定义如下:

```
struct books{
    char booknum[20];                /*书号*/
    char bookname[20];               /*书名*/
    char authorname[20];             /*作者*/
    char cbs[20];                    /*出版社*/
    char price[5];                   /*价格*/
    struct books * next;
    struct books * prior;
};
```

(2) 对于图书信息的各种操作如添加、浏览、查询、删除等都用链表来处理,需要定义两个分别指向图书信息链表头结点和尾结点的全局指针变量:

```
struct books * head;                 /*头结点*/
struct books * last;                 /*尾结点*/
```

(3) 为实现本系统的各项功能,需定义函数如下:

```
int menu(void);                                                  /*菜单选项*/
void inputs(char * prompt,char * s,int count);                   /*输入格式*/
void enter(void);                                                /*信息输入*/
void dls_store(struct books * i,struct books **head,struct books **last);
                                                                 /*按书号大小建立链表*/
struct books * find(char * bn);                                  /*按书名检索*/
void search(void);                                               /*按书名查找*/
void modify(void);                                               /*修改信息*/
void display(struct books * p);                                  /*显示格式*/
void list(void);                                                 /*将图书信息详细地列出来*/
void del(struct books **head,struct books **last);               /*删除信息*/
void save(void);                                                 /*以文件形式保存信息*/
void load(void);                                                 /*从文件中载入信息*/
```

(4) 将整个系统分为以下几个模块来实现:添加图书信息模块、浏览图书信息模块、查询图书信息模块(分别按不同条件进行查询)、修改图书信息模块、删除图书信息模块以及图书信息保存模块。

根据以上算法分析,该系统的程序主框架功能结构如图 7.1 所示。

图 7.1　程序主框架结构图

程序流程图如图 7.2 所示。

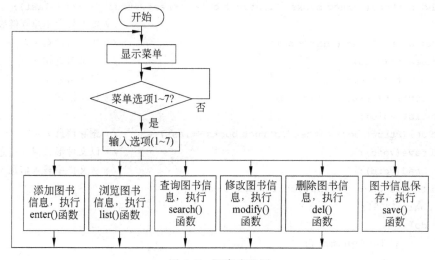

图 7.2　程序流程图

2. 难点提示

(1) 首先要创建链表,并可以从文件数据库中读取已经备份的图书信息(若文件已存储图书信息)到链表中。可以使用不包含空头结点的链表,就是在头结点里存储了一条图书信息的链表。

(2) 设计对链表相关操作的函数:包括插入一个结点,根据 ID 查找结点,删除一个结点等。添加时,函数应该将输入的图书信息按书号的顺序插入到链表中,查询时首先利用循环遍历链表中的每一个结点,直到查到符合条件的为止。

(3) 在读取文件中图书信息到链表中和将整个链表中存储的图书信息保存到文件中,可使用 fread 和 fwrite 函数来将结构体中的数据存取于文件。

3. 程序代码

```
#include<stdio.h>
#include<stdlib.h>
#include<string.h>
struct books{
    char booknum[20];                    /*书号*/
```

```c
        char bookname[20];                              /*书名*/
        char authorname[20];                            /*作者*/
        char cbs[20];                                   /*出版社*/
        char price[5];                                  /*价格*/
        struct books * next;
        struct books * prior;
};
struct books * head;                                    /*头结点*/
struct books * last;                                    /*尾结点*/
int menu(void);                                         /*菜单选项*/
void inputs(char * prompt,char * s,int count);          /*输入格式*/
void enter(void);                                       /*信息输入*/
void dls_store(struct books * i,struct books **head,struct books **last);
                                                        /*按书号大小建立链表*/
struct books * find(char * bn);                         /*按书名检索*/
void search(void);                                      /*按书名查找*/
void modify(void);                                      /*修改信息*/
void display(struct books * p);                         /*显示格式*/
void list(void);
void del(struct books **head,struct books **last);      /*删除信息*/
void save(void);                                        /*以文件形式保存信息*/
void load(void);                                        /*从文件中载入信息*/
void inputs(char * prompt,char * s,int count)
{   char p[255];
     do{
         printf(prompt);
         gets(p);
         if(strlen(p)>count)                            /*限定输入字符长度*/
            printf("\nToo long\n");
     }while(strlen(p)>count);
     strcpy(s,p);
}
void enter(void)                                        /*输入信息*/
{   char ch;
     struct books * p;
     while(1)
     {   p=(struct books *)malloc(sizeof(struct books));
         inputs("请输入书的编号:",p->booknum,20);
         inputs("请输入书名:",p->bookname,20);
         inputs("请输作者名:",p->authorname,20);
         inputs("请输入出版社:",p->cbs,20);
         inputs("请输入图书价格:",p->price,5);
         printf("\n");
         dls_store(p,&head,&last);
         fflush(stdin);
```

```c
            printf("是否继续输入(Y/N):");
            ch=getchar();
            fflush(stdin);
            if(ch=='N'||ch=='n') break;
        }
}
void dls_store(struct books * i,struct books **head,struct books **last)
                                                        /*按书号顺序插入表中
{   struct books * p, * old;
    if((* last)==NULL)
    {   i->prior=NULL;
        i->next=NULL;
        (* head)=i;
        (* last)=i;
        return;
    }
    p=(struct books * )malloc(sizeof(struct books));
    p=(* head);
    old=NULL;
    while(p!=NULL)
    {   if(strcmp(p->booknum,i->booknum)<0)
        {   old=p;    p=p->next;   }
        else
        {   if(p->prior)
            {   p->prior->next=i;
                i->next=p;
                i->prior=p->prior;
                p->prior=i;
                return;
            }
            i->next=p;                                  /*放表头*/
            i->prior=NULL;
            p->prior=i;
            * head=i;
            return;
        }
    }
    old->next=i;
    i->next=NULL;
    i->prior=old;
    * last=i;
}
struct books * find(char * bn)                          /*按书名搜索关键字*/
{   struct books * p;
    p=head;
```

```c
        while(p!=NULL)
        {   if(strcmp(bn,p->bookname)==0)    return p;
            p=p->next;
        }
        return NULL;
}
void search(void)
{   char name[20];
    struct books  * p;
    printf("请输入你要查询的书名:");
    gets(name);
    p=find(name);
    if(p==NULL)   printf("没有找到你查询的信息.\n");
    else display(p);
}
void modify(void)                                       /*根据书名查找并修改信息*/
{   char name[20];
    struct books  * p;
    printf("请输入你要查询的书名:");
    gets(name);
    p=find(name);
    if(p==NULL)   printf("没有找到你查询的信息.\n");
    else
    {   inputs("请输入书的编号:",p->booknum,20);
        inputs("请输入书名:",p->bookname,20);
        inputs("请输作者名:",p->authorname,20);
        inputs("请输入出版社:",p->cbs,20);
        inputs("请输入图书价格:",p->price,5);
        printf("\n");
    }
}
void display(struct books * p)
{   printf("%s\t",p->booknum);                          /*书号*/
    printf("%s\t",p->bookname);                         /*书名*/
    printf("%s\t",p->authorname);                       /*作者*/
    printf("%s\t",p->cbs);                              /*出版社*/
    printf("%s\t",p->price);
    printf("\n");
}
void list(void)
{                                                       /*system("cls");
    struct books  * p;
    p=head;
    printf("书号\t 书名\t 作者\t 出版社\t 价格\n");
    while(p !=NULL)
```

```c
        { display(p);p=p->next; }
    printf("\n");
}
void del(struct books * * head,struct books * * last)
{   struct books * i;
    char s[50];
    inputs("输入书名:",s,20);
    i=find(s);
    if(i)
    {   if(* head==i)
        {   * head=i->next;
            if(* head) (* head)->prior=NULL;
            else  * last=NULL;
        }
        else
        {   i->prior->next=i->next;
            if(i!=* last)i->next->prior=i->prior;
            else  * last=i->prior;
        }
        free(i);
    }
}
void save(void)
{   struct books * p;
    FILE * fp;
    if((fp=fopen("books","wb")==NULL)
    {   printf("open file error.\n"); exit(1); }
    p=head;
    while(p!=NULL)
    {   fwrite(p,sizeof(struct books),1,fp);
        p=p->next;
    }
    fclose(fp);
}
void load(void)
{   struct books * p;
    FILE * fp;
    fp=fopen("books","rb");
    if(fp==NULL)
    {   printf("open file error.\n");    exit(1); }
    while(head)
    {   p=head->next;
        free(p);
        head=p;
    }
```

```c
        head=last=NULL;
        printf("\n 载入信息 \n");
        while(!feof(fp))
        {   p=(struct books *)malloc(sizeof(struct books));
            if(1!=fread(p,sizeof(struct books),1,fp)) break;
            dls_store(p,&head,&last);
        }
        fclose(fp);
}
int menu(void)
{   char s[20];
    int c;
    printf("\t\t*****************图书信息管理系统*****************\n");
    printf("\t\t * \t\t1.图书信息输入                    * \n");
    printf("\t\t * \t\t2.图书信息浏览                    * \n");
    printf("\t\t * \t\t3.图书信息查询                    * \n");
    printf("\t\t * \t\t4.图书信息修改                    * \n");
    printf("\t\t * \t\t5.图书信息删除                    * \n");
    printf("\t\t * \t\t6.存储图书信息                    * \n");
    printf("\t\t * \t\t7.载入图书信息                    * \n");
    printf("\t\t * \t\t8.退出系统                        * \n");
    printf("\t\t*****************图书信息管理系统*****************\n\n\n");
    printf("\t\t 请输入以上序号进行选择:");
    do{  gets(s);
         c=atoi(s);
    }while(c<0||c>8);
    return c;
}
int main(void)
{   head=last=NULL;
    while(1)
    {   switch(menu())
        {   case 1:enter();                         /*录入*/
                break;
            case 2:list();                          /*浏览*/
                break;
            case 3:search();                        /*查找*/
                break;
            case 4:modify();                        /*修改*/
                break;
            case 5:del(&head,&last);                /*删除*/
                break;
            case 6:save();                          /*保存*/
                break;
            case 7:load();                          /*装载*/
```

                break;
            case 8:exit(0);
        }
    }
    return 0;
}

4. 运行结果示例

(1) 程序运行时,显示功能项菜单如图 7.3 所示。

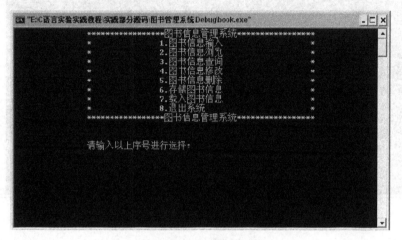

图 7.3　显示功能项菜单

(2) 输入功能项菜单 1,录入图书信息,可以循环录入多个图书的信息。当停止录入图书信息时,输入 N 或 n 结束,如图 7.4 所示。

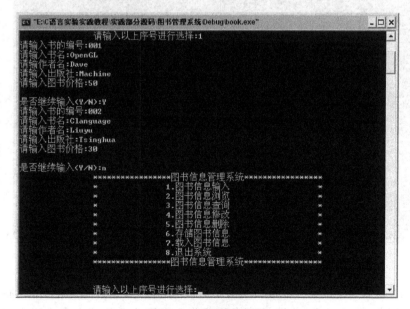

图 7.4　录入图书信息

(3) 选择功能项菜单 6，将以上录入的信息存入到磁盘文件中，成功存储文件后，选择功能项菜单 7 载入图书信息后，选择功能项菜单 2 可进行图书信息的浏览，如图 7.5 所示。

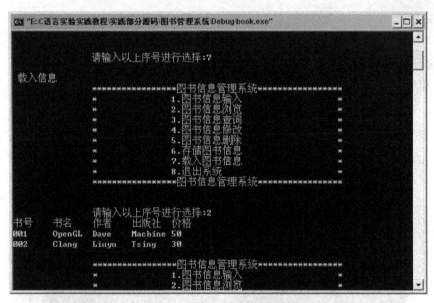

图 7.5　存储并载入图书信息

(4) 选择功能项菜单 3 可进行图书信息的查询，可按书名进行查询，如图 7.6 所示。

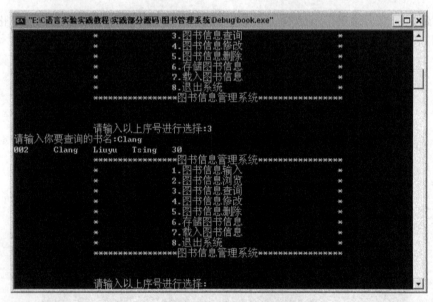

图 7.6　图书信息查询

(5) 选择功能项菜单 4 可进行图书信息的修改，输入要修改书的名称后，接下来修改该书的各项信息，输入完毕后，选择功能项菜单 6 将修改后的信息存入磁盘文件。接下来选择菜单项 2 浏览图书信息，可以看到该条图书信息已成功修改，分别如图 7.7 和

图7.8所示。

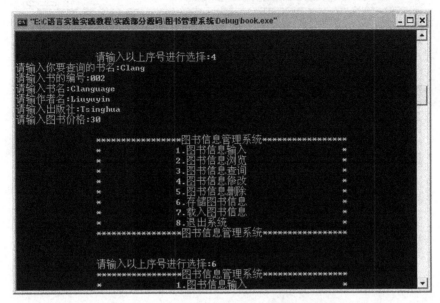

图7.7 修改后的信息存入磁盘文件

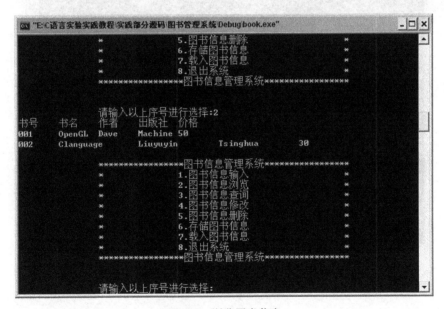

图7.8 浏览图书信息

(6) 选择功能项菜单5可进行图书信息的删除,删除结束后,选择功能项菜单6存储图书信息,然后再选择功能菜单2进行图书信息浏览时,可以看到该条记录已不存在,分别如图7.9和图7.10所示。

(7) 选择功能菜单8退出系统。

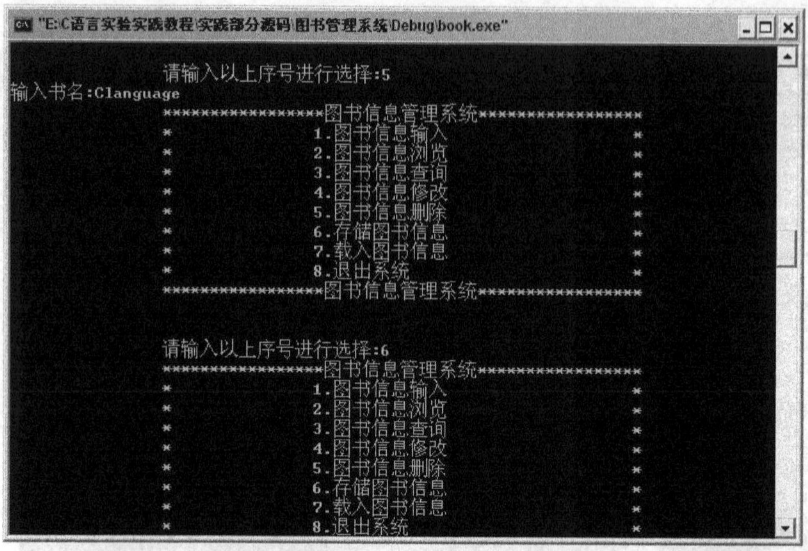

图 7.9 图书信息删除

图 7.10 图书信息删除后

项目八 贪吃蛇游戏

本项目要求设计一个简单的贪吃蛇游戏。游戏规则是:玩家控制一条蛇来吃屏幕上随机产生的食物,通过上下左右箭头来控制蛇的移动方向,如果蛇碰到周围的边框,则游戏结束。游戏过程中需要在屏幕上显示玩家的得分,每吃掉一个食物得 10 分。

1. 算法分析

(1) 设计本游戏有两个关键点:一个是如何表示蛇以及食物对象,另外一个是怎样来控制蛇的移动。

(2) 为简单起见,此游戏采用一个绿色矩形块来表示食物,一个红色矩形块来表示蛇的一节身体,蛇头使用两节来表示,每吃到一个食物蛇身会增加一节。表示食物和蛇的矩形块都设计为 10×10 个像素。程序中定义如下的数据结构来表示食物和蛇。

```
/*定义食物的结构体*/
struct Food
{   int x;                          /*食物的横坐标*/
    int y;                          /*食物的纵坐标*/
    int need;                       /*判断是否要出现食物的变量*/
};
/*定义蛇的结构体*/
struct Snake
{   int x[NODE];
    int y[NODE];
    int node;                       /*蛇的节数*/
    int direction;                  /*蛇移动方向*/
    int life;                       /*蛇的生命,0 活着,1 死亡*/
};
```

2. 难点提示

(1) 在食物的结构体 Food 中,通过(x,y)来确定它的坐标位置,可以采用随机数的方式来确定一个食物的位置。成员 need 用来确定是否应该在屏幕上生成一个食物。在 Snake 结构体中,通过(x[i],y[i])来确定第 i 节蛇身的坐标位置,NODE 是事先定义好的常量,它指定蛇的最大身长,成员变量 node 则记录当前状态下蛇的身长,成员变量 direction 和 life 分别表示蛇的移动方向和生命,如果 life 为 0 则表示蛇活着,可以继续移动吃食物,如果 life 为 1 则表示蛇死亡,游戏结束。

(2) 因为程序中蛇的身长是不断增长的,所以需要实时地绘制一个变化的蛇,可以定义一个函数来实现这个功能。

(3) 可以定义一个函数来具体实现蛇的移动以及整个游戏的控制过程。

3. 程序代码

```
#include<graphics.h>
```

```c
#include<stdlib.h>
#include<dos.h>
/*定义控制命令*/
#define LEFT   0x4b00
#define RIGHT  0x4d00
#define DOWN   0x5000
#define UP     0x4800
#define ESC    0x011b
/*定义蛇的最大节数*/
#define NODE 200
int i,key;
int score=0;                             /*得分*/
/*定义游戏速度*/
int SPEED=100;
/*定义食物的结构体*/
struct Food
{  int x;                                /*食物的横坐标*/
   int y;                                /*食物的纵坐标*/
   int need;                             /*判断是否要出现食物的变量*/
};
/*定义蛇的结构体*/
struct Snake
{  int x[NODE];
   int y[NODE];
   int node;                             /*蛇的节数*/
   int direction;                        /*蛇移动方向*/
   int life;                             /*蛇的生命,0活着,1死亡*/
};
struct Food food;
struct Snake snake;
/*开始画面,左上角坐标为(50,40),右下角坐标为(600,450)的围墙*/
void DrawFence(void)
{  setcolor(WHITE);
   /*画围墙的上边*/
   line(50,40,600,40);
   line(50,42,600,42);
   /*画围墙的下边*/
   line(50,450,600,450);
   line(50,452,600,452);
   /*画围墙的左边*/
   line(50,40,50,450);
   line(52,40,52,450);
   /*画围墙的右边*/
   line(600,40,600,450);
```

```c
    line(602,40,602,450);
}
/*输出成绩*/
void PrintScore()
{   char string[10];
    setfillstyle(SOLID_FILL,YELLOW);
    bar(50,15,220,35);
    setcolor(RED);
    settextstyle(0,0,2);
    sprintf(string,"score:%d",score);
    outtextxy(55,20,string);
}
/*游戏结束*/
void GameOver()
{   cleardevice();
    PrintScore();
    setcolor(RED);
    settextstyle(0,0,4);
    outtextxy(200,200,"GAME OVER");
    getch();
}
/*画蛇函数*/
void DrawSnake()
{   setcolor(4);
    for(i=0;i<snake.node;i++)
    rectangle(snake.x[i],snake.y[i],snake.x[i]+10,snake.y[i]-10);
    delay(SPEED);
    setcolor(0);
    rectangle(snake.x[snake.node-1],snake.y[snake.node-1],
        snake.x[snake.node-1]+10,snake.y[snake.node-1]-10);
}
/*玩游戏具体过程*/
void PlayGame()
{   randomize();                    /*随机数发生器*/
    food.need=1;                    /*1表示需要出现新食物,0表示已经存在食物*/
    snake.life=0;                   /*活着*/
    /*方向往右*/
    snake.direction=1;
    /*蛇头*/
    snake.x[0]=100;
    snake.y[0]=100;
    snake.x[1]=110;
    snake.y[1]=100;
    snake.node=2;                   /*节数*/
```

```c
PrintScore();                           /*输出得分*/
while(1)                                /*可以重复玩游戏,压 Esc 键结束*/
{   while(!kbhit())                     /*在没有按键的情况下,蛇自己移动身体*/
    {   if(food.need==1)                /*需要出现新食物*/
        {   food.x=rand()%400+60;
            food.y=rand()%350+60;
            while(food.x%10!=0)
                                        /*食物随机出现后必须让食物能够在整格内,这样才可以让蛇吃到*/
            food.x++;
            while(food.y%10!=0)food.y++;
            food.need=0;                /*画面上有食物了*/
        }
        if(food.need==0)                /*画面上有食物了就要显示*/
        {   setcolor(GREEN);
            rectangle(food.x,food.y,food.x+10,food.y-10);
        }
        for(i=snake.node-1;i>0;i--)
        /*蛇的每个环节往前移动,也就是贪吃蛇的关键算法*/
        {   snake.x[i]=snake.x[i-1];
            snake.y[i]=snake.y[i-1];
        }
        /*1,2,3,4 表示右,左,上,下四个方向,通过这个判断来移动蛇头*/
        switch(snake.direction)
        {   case 1:snake.x[0]+=10;break;
            case 2: snake.x[0]-=10;break;
            case 3: snake.y[0]-=10;break;
            case 4: snake.y[0]+=10;break;
        }
        for(i=3;i<snake.node;i++)
        /*从第四节开始判断是否撞到自己了,因为蛇头为两节,第三节不可能拐过来*/
            if(snake.x[i]==snake.x[0]&&snake.y[i]==snake.y[0])
        GameOver();                     /*显示失败*/
        snake.life=1;
        break;
    }
    if(snake.x[0]<55||snake.x[0]>595||snake.y[0]<55||snake.y[0]>455)
                                        /*蛇是否撞到墙壁*/
    {   GameOver();                     /*本次游戏结束*/
        snake.life=1;                   /*蛇死*/
    }
    if(snake.life==1)                   /*以上两种判断以后,如果蛇死就跳出内循环,重新开始*/
        break;
    if(snake.x[0]==food.x&&snake.y[0]==food.y)         /*吃到食物以后*/
```

```
            {    setcolor(0);                           /*把画面上的食物东西去掉*/
                 rectangle(food.x,food.y,food.x+10,food.y-10);
                 snake.x[snake.node]=-20;snake.y[snake.node]=-20;
                               /*新的一节先放在看不见的位置,下次循环就取前一节的位置*/
                 snake.node++;                  /*蛇的身体长一节*/
                 food.need=1;                   /*画面上需要出现新的食物*/
                 score+=10;
                 PrintScore();                  /*输出新得分*/
            }
            DrawSnake();
        }
        if(snake.life==1) break;             /*如果蛇死就跳出循环*/
        key=bioskey(0);                      /*接受按键*/
        if(key==Esc) break;                  /*按 Esc 键退出*/
        else if(key==UP&&snake.direction!=4)
                    /*判断是否往相反的方向移动*/
        snake.direction=3;
        else if(key==RIGHT&&snake.direction!=2)
                 snake.direction=1;
        else if(key==LEFT&&snake.direction!=1)
                 snake.direction=2;
        else if(key==DOWN&&snake.direction!=3)
                 snake.direction=4;
    }
}
/*主函数*/
void main(z)
{   int gd=DETECT,gm;
    initgraph(&gd,&gm,"e:\\tc");
    cleardevice();
    /*绘制边框*/
    DrawFence();
    /*玩游戏具体过程*/
    PlayGame();
    getch();
    closegraph();
}
```

4. 运行结果示例

(1) 程序开始运行时,蛇从左上部出来,蛇身两节,食物的位置随机产生,如图 8.1 所示。

(2) 通过上下左右方向键调整蛇的运动方向,让它朝食物所在地运动,当蛇头部运动到食物所在地时,分数会加上 10 分,同时蛇身会长出一节,如图 8.2～图 8.4 所示。

(3) 蛇运行过程中,当蛇头碰到边框时,游戏结束,结束画面与前面俄罗斯游戏类似。

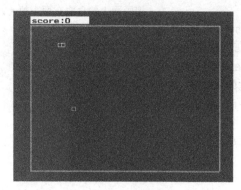
图 8.1 游戏开始运行

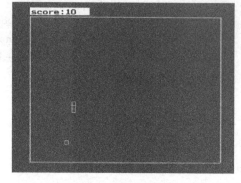
图 8.2 蛇长到三节

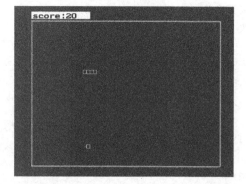
图 8.3 蛇长到四节

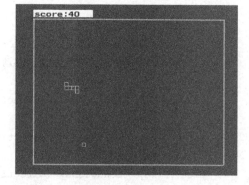
图 8.4 蛇长到六节

第三部分

习题参考答案

第三部分

刘大櫆考论

第1章 C程序知识初步

1. 合法的变量名有：Abc,s_c,Float,w5,CHAR,begin,a2b
2. 填空题
(1) scanf() printf()
(2) main(主)
(3) /* */
(4) 分号(;)
(5) 函数
(6) 英文字母(大写或小写) 数字 下划线 数字 不同
(7) 错误/error 警告/warnning
3.

```
#include<stdio.h>
void main()
{   printf("学号    姓名    年龄    性别    专业    班级\n");
    printf("%s  %s  %d  %c  %s  %s\n","110112","李林",19,'M',"计算机","001");
}
```

4. 将案例1.4中循环变量i的起始值改为1即可。

```
#include<stdio.h>
void main()
{   int i,sum=0;
    for(i=1;i<=100;i=i+2)
        sum=sum+i;
    printf("1~100之间的偶数和为:%d\n",sum);
}
```

5. 交换两种饮料的瓶子,必须借助另一个空瓶。可先将橙汁倒进空瓶,将苹果汁倒进原来盛橙汁的瓶子,再将橙汁倒进原来盛苹果汁的瓶子即可。也可以是相反的次序。

第 2 章 基本数据类型及其操作

1. 判断题
√×× √× ×× √√ √

2. 填空题
(1) int a;float b;char c;
(2) 3
(3) double
(4) a=1,b=2,c=3
(5) 0 整型(int)
(6) 2
(7) 1
(8) f

3. 选择题
ACDDB DBACD

4. 改错题
(1)

#include<stdoi.h>   改成   #include<stdio.h>
int celsius;fahr;   改成   float celsius,fahr;
printf("fahr=d,celsius=%d\n",fahr,celsius); 改成 printf("fahr=%f,celsius=%f\n",fahr,celsius);

(2)

printf("%d\n",m/n+a);   改成   printf("%f\n",m/n+a);
printf("%f\n",m%a+b);   改成   printf("%f\n",m%(int)a+b);

(3) 下面程序中有两处错误，请查找并改正。

#include<stdio.h>
void main()
{   int m,n;
    float a,b;
    scanf("%d,%f",&m,&a);
    scanf("%d,%f",&b,&n);   改成   scanf("%f,%d",&b,&n);
    printf("%f\n",m/n+a);
    printf("%f\n",m%n+b);   改成   printf("%f\n",m%n+b);
}

5. 阅读程序
(1) 2

(2)
```
    *
   * *
  *   *
   * *
    *
```
(3) 程序的功能：从键盘任意输入两个整数，输出它们的和。
(4) 26.765432
6. 编程题
(1)
```
#include<stdio.h>
void main()
{   float celsius,fahr;
    fahr=50;
    celsius=5*(fahr-32)/9;
    printf("fahr=%f,celsius=%f\n",fahr,celsius);
}
```

(2)
```
#include<stdio.h>
void main()
{   int a,b;
    scanf("%d",&a);
    b=a*a;
    printf("a=%d,b=%d\n",a,b);
}
```

(3)
```
#include<stdio.h>
void main()
{   int x,y,z;
    scanf("%d%d",&x,&y);
    z=x%y;
    printf("%d对%d的余数为%d\n",x,y,z);
}
```

(4)
```
#include<stdio.h>
void main()
{   int x,y,t;
    x=3;  y=4;
    t=x;  x=y;  y=t;
    printf("x=%d,y=%d \n",x,y);
```

}

(5)

```c
#include<stdio.h>
void main()
{   int n,x,y,z;
    n=152;
    x=n/100;
    y=n/10%10;
    z=n%10;
    printf("%d 的百位、十位、个位分别为%d,%d,%d\n",n,x,y,z);
}
```

第3章 选择结构程序设计

1. 选择题
CBCCC　DCADA
2. 填空题
(1) 10.500000
(2) 优先级　结合性
(3) ASCII 码
(4) 0.000000
(5) 0　1
(6) (ch>='a'&&ch<='z') || (ch>='A'&& ch<='Z') || (ch>='0'&&ch<='9')
(7) ① c=getchar()　　② %c　　　③ &c　　④ putchar(c)
(8) ① t/3　　　　　　② default　　③ 1-d/100.0
3. 分析下列程序，写出程序运行结果。
(1) 123.456001** 123.46**123.456
(2) 如果输入 3 和 4，则输出结果为 max=4
(3) -1
(4) 5
(5) a=2,b=1
4. 改错题
(1)

scanf("%c, %d, %d, %f ",c, b, a, c); 改成　scanf("%c, %f, %d ",&c, &b, &a);
printf("%c, %d, %d, %f ",c, b, a, c); 改成　printf("%c, %f, %d, %d ",c, b, a, c);

(2)

if(x>y) x=y;y=x;　　　改成　if(x>y) { x=y;y=x; }
else x=x+1;y=y+1;　　改成　else { x=x+1;y=y+1; }

5. 编程题
(1)

```
#include<stdio.h>
void main()
{   float x,y;
    scanf("%f",&x);
    if(x<=15)  y=4*x/3;
    else y=2.5*x-10.5;
    printf("y=f(%f)=%f",x,y);
```

}

(2)

```
#include<stdio.h>
#include<math.h>
void main()
{   int t;
    scanf("%d",&t);
    if (t%2==1)   printf("%d 的平方根为%f",t,sqrt(t));
    else printf("%d 的立方根为%f",t,pow(t,1.0/3));
}
```

(3)

```
#include<stdio.h>
void main()
{   int t;
    scanf("%d",&t);
    if(t%3==0 && t%7==0) printf("%d 能同时被 3 和 7 整除\n",t);
    else printf("%d 不能同时被 3 和 7 整除\n",t);
}
```

(4)

```
#include<stdio.h>
void main()
{   char ch;
    ch=getchar();
    if(ch>='a' && ch<='z')printf("%c 是小写字母\n",ch);
    else printf("%c 是大写字母\n",ch);
}
```

(5)

```
#include<stdio.h>
void main()
{   char ch;
    ch=getchar();
    if(ch>='a' && ch<='z') printf("%c\n",ch);
    else if (ch>='A' && ch<='Z')   printf("%c\n",ch+32);
    else printf("%c 不是英文字母\n",ch);
}
```

(6)

```
#include<stdio.h>
void main()
{   int a,b,c,max;
    scanf("%d%d%d",&a,&b,&c);
```

```
        if(a>b)max=a;
        else max=b;
        if (max<c)max=c;
        printf("%d,%d,%d 中的最大数是%d\n",a,b,c,max);
}
```

(7)

```
#include<stdio.h>
void main()
{   float x,y;
    scanf("%f",&x);
    if(x<-10)   y=10;
    else if(x<=10)   y=0;
    else y=-10;
    printf("y=f(%f)=%f",x,y);
}
```

(8)

```
#include<stdio.h>
void main()
{   int a,b,c,d,t;
    scanf("%d%d%d%d",&a,&b,&c,&d);
    /*先找出 4 个数中最大数,并存放在变量 a 中*/
    if(a<b) {t=a;    a=b;    b=t;}
    if(a<c) {t=a;    a=c;    c=t;}
    if(a<d) {t=a;    a=d;    d=t;}
    //找出除 a 外其余 3 个数中的最大数,并存放在变量 b 中
    if(b<c) {t=b;    b=c;    c=t;}
    if(b<d) {t=b;    b=d;    d=t;}
    //找出变量 c、d 中的大数,并存放在变量 c 中
    if(c<d) {t=c;    c=d;    d=t;}
    printf("%d %d %d %d\n",a,b,c,d);
}
```

(9)

```
#include<stdio.h>
void main()
{   int year,day,t;
    scanf("%d%d",&year,&day);
    switch(day)
    {   case 1:case 3:case 5:case 7:case 8:
             case 10:case 12:t=31;break;
        case 2:if((year%400==0)||(year%4==0&&year%100!=0))t=29;
             else t=28; break;
        default:t=30;
```

```
    }
    printf("%d年%d月有%d天\n",year,day,t);
}
```

(10)

```c
#include<stdio.h>
void main ()
{   float salary,tax;
    int rate;
    scanf("%f",&salary);
    if(salary<=2500) rate=0;
    else if(salary>2500  &&  salary<=4000) rate=5;
    else if(salary>4000  &&  salary<=7000) rate=10;
    else if(salary>7000  &&  salary<=22000) rate=15;
    else if(salary>22000 &&  salary<=42000) rate=20;
       else if(salary>42000 &&  salary<=62000) rate=25;
    else if(salary>62000 &&  salary<=82000) rate=30;
    else if(salary>82000 &&  salary<=102000) rate=35;
    else   rate=40;
    tax=rate * 0.01 * (salary-2000);
    printf("%f\n",tax);
}
```

第4章 循环结构

1. 选择题
D(CAA)AB(AD)　　BBCCB
2. 填空题
(1) 0
(2) －264
(3) 354％10　　354/100　　354/10％10
(4) 提前结束整个循环　　提前结束本次循环
(5) n<＝999　　ns＝n/10％10　　ng＋ns＋nb＝＝5
(6) ① sum＝0　　② sum　　③ b＋2
3. 写出下列程序的运行结果
(1) x＝0，n＝5
(2) i＝6,y＝10
(3) sum＝288
(4) K＝4
(5) 1.600000
4. 编程题
(1)

```
#include<stdio.h>
void main()
{   double a,sum=0;
    int i,m=1,n,j;
    printf("输入一个数：");
    scanf("%d",&n);
    for(i=1;i<=n;i++)
    {  if(i%2==0)m=-1;
       else m=1;
       a=2*i-1;
       sum=sum+m*1.0/a;
    }
    printf("%lf\n",sum);
}
```

(2)

```
#include<stdio.h>
void main()
{  int a,b=0;
```

```
    for(a=1;a<=100;a++)
    {  if(a%2==0)
        {  printf("%6d",a),
           b++;
           if(b%6==0)printf("\n");
        }
    }
}
```

(3)

```
#include<stdio.h>
void main()
{   int j,n,t;
    scanf("%d",&n);
    t=1;
    for(j=1;j<=n;j++) t=t*j;
    printf("%d\n",t);
}
```

(4)

```
#include<stdio.h>
#include<math.h>
void main()
{   double x;
    int n;
    scanf("%lf",&x);
    n=(int)(x*x);
    printf("%d",n);
}
```

(5)

```
#include<stdio.h>
#include<math.h>
void main()
{   int t,i=0,s;
    scanf("%d",&t);
    s=t;   while((s=s/10)!=0) i++;
    do{
        printf("%d ",t/(int)pow(10,i));
        t=t%(int)pow(10,i);
        i--;
    }while(i>=0);
}
```

(6)

```c
#include<stdio.h>
void main()
{   int n,min,i,x;
    printf("请输入一个整数 n:\n");
    scanf("%d",&n);
    printf("再输入 n 个整数：\n");
    for(i=1;i<=n;i++)
    {   scanf("%d",&x);
        if(i==1) min=x;
        else if(x<min)min=x;
    }
    printf("min=%d\n",min);
}
```

(7)

```c
#include<stdio.h>
void main()
{   int i,j;
    for(i=1;i<=9;i++)
    {   for(j=1;j<=i;j++)
            printf("%d*%d=%-3d",j,i,i*j);
        printf("\n");
    }
}
```

(8)

```c
#include<stdio.h>
void main()
{ int i,j,k,t=3;
  for(i=1;i<=4;i++)
  {   for(j=1;j<=t;j++)printf(" ");
      t--;
      for(j=1;j<=2*i-1;j++)printf("* ");
      printf("\n");
  }                           /*上半部分图形*/
  t=5;
  for(i=1;i<=3;i++)
  {   for(j=1;j<=i;j++)printf(" ");
      for(j=1;j<=t;j++)printf("* ");
      t=t-2;
      printf("\n");
  }
}
```

(9)

```c
#include<stdio.h>
void main()
{   int i,a,n,t,s=0;
    scanf("%d%d",&a,&n);
    t=a;
    for(i=1;i<=n;i++)
    {   s=s+t;
        t=t*10+a;
    }
    printf("s=%d\n",s);
}
```

(10)

```c
#include<stdio.h>
void main()
{   float h,sum;
    int i;
    h=100;   sum=100;
    for(i=1;i<=10;i++)
    {   h=h/2.0;
        sum=sum+2*h;
    }
    sum=sum-2*h;
    printf("第%d次弹起的高度为%f米,经过的距离为%f米\n",i-1,h,sum);
}
```

(11)

```c
#include<stdio.h>
void main()
{   int i,x,y;      /*x表示某一天没吃之前桃子的数,y表示该天吃过之后剩下桃子的数*/
    y=1;i=9;        /*第9天吃过桃子之后,还剩下 y=1个桃子*/
    while(i>=1)
    {   x=2*(y+1);
        y=x;
        i--;
    }
    printf("%d\n",x);
}
```

第 5 章 数 组

1. 判断题
TFTFF TTF
2. 选择题
CCACA BDADD
3. 分析以下程序的运行结果
(1) 5,6.000000
(2) 3,5,7,9,5
(3) Cat
(4) aaaa
    bb
    cc
(5) str=Language
(6) 123
(7) str[]=abdef
(8) 2,3
4. 编程题
(1)

```c
#include<stdio.h>
void main()
{   int z=0,l=0,f=0,i,a[6]={8,0,9,0,-6,-1};
    for(i=0 ;i<6;i++)
    {   if(a[i]==0)l++;
        else if(a[i]>0)z++;
        else if(a[i]<0)f++;
    }
    printf("数组中正数的个数为%d,负数的个数为%d,零的个数为%d\n",z,f,l);
}
```

(2)

```c
#include<stdio.h>
void main()
{   int i,score[50];
    int highest,lowest,average=0;
    for(i=0;i<50;i++)
        scanf("%d",&score[i]);
    highest=score[0];
```

```
        lowest=score[0];
        for(i=0;i<50;i++)
        {   if(score[i]>highest) highest=score[i];
            if(score[i]<lowest)lowest=score[i];
            average+=score[i];
        }
        average=average/50;
        printf("%d, %d, %d\n",highest,lowest,average);}
```

(3)

```
#include<stdio.h>
int main()
{   int i;
    int f[20]={1,1};              /*f[0]=1,f[1]=1*/
    for(i=2;i<20;i++)
        f[i]=f[i-2]+f[i-1];
    for(i=0;i<20;i++)             /*此循环的作用是输出20个数*/
    {   if(i%5==0) printf("\n");  /*控制换行,每行输出5个数据*/
        printf("%8d",f[i]);       /*每个数据输出时占8列宽度*/
    }
    return 0;
}
```

(4)

```
#include<stdio.h>
#include<string.h>
int fun(char str[],char sub[])
{   int i,n=0;
    for(i=0;i<strlen(str);i++)
    if((str[i]==sub[0])&&(str[i+1]==sub[1]))n++;
    return n;
}
void main()
{   char str[80],sub[3];
    int n;
    printf("请输入主字符串：\n");
    gets(str);
    printf("请输入子字符串：\n");
    gets(sub);
    n=fun(str,sub);
    printf("子字符串在主字符串中出现的次数为%d\n",n);
}
```

(5)

```
#include<stdio.h>
```

```
#include<string.h>
void main()
{   int m,n;
    char str1[80],str2[80];
    printf("请输入两个字符串\n");
    gets(str1);
    gets(str2);
    m=strlen(str1);
    n=strlen(str2);
    printf("str1 长度为:%d,str2 长度为:%d\n",m,n);
    if(strcmp(str1,str2)==0) printf("两个字符串相同!\n");
    else printf("两个字符串不相同!\n");
    printf("连接后 str1 为:%s\n",strcat(str1,str2));
}
```

(6)

```
#include<stdio.h>
#include<string.h>
void main()
{   char str[80];
    int i,n;
    printf("请输入一个字符串\n");
    gets(str);
    for(i=0;str[i]!='\0';i++)
    {   if(str[i]>='a'&&str[i]<='y'){   str[i]=str[i]+1;   }
        else   if(str[i]=='z') str[i]='a';
    }
    printf("%s\n",str);
}
```

(7)

```
#include<stdio.h>
#include<string.h>
#define MAX 5
void main()
{   char str[MAX][50];
    char s[50];
    int i=0;
    gets(str[i]);
    strcpy(s,str[i]);
    for(i=1;i<MAX;i++)
    {   gets(str[i]);
        if((strcmp(str[i],s)>0))strcpy(s,str[i]);
    }
    printf("最大的串为:%s\n",s);
```

}

(8)

```c
#include<stdio.h>
#define N 50
void main()
{   int primes[N];
    int i,m,k,n;
    printf("The first %d prime numbers are:\n",N);
    primes[0]=2;                /* 2是第一个质数 */
    i=1;                        /* 已有第一个质数 */
    m=3;                        /* 被测试的数从 3 开始 */
    while(i<N)
    {   k=0;
        while(primes[k]*primes[k]<=m)
          if(m%primes[k]==0)
            {   /* m 是合数 */
               m+=2;            /* 让 m 取下一个奇数 */
               k=1;             /* 不必用 primes[0]=2 去测试 m,所以 k 从 1 开始 */
            }
          else   k++;           /* 继续用下一个质数去测试 */
        primes[i++]=m;
        m+=2;                   /* 除 2 外,其余质数均是奇数 */
    }
    for(k=0;k<i;k++)            /* 输出 primes[0]至 primes[pc-1] */
       printf("%4d",primes[k]);
    printf("\n请输入一个整数：");
    scanf("%d",&n);
    for(i=0;i<N/2;i++)
       if(n%primes[i]==0)break;
    if(i>=N/2)printf("%d is prime!\n",n);
    else printf("%d is not prime!\n",n);
}
```

第6章 函　　数

1. 判断下面叙述的对与错
FFFFF　TFTFF
2. 选择题
DAACC　ABDDB
3. 分析以下程序的运行结果
(1) 3.0
(2) 6
(3) 8
(4) 7
(5) x=2,y=3
(6) 4
(7) 817
4. 编程题
(1)

```
int shuixianhua()
{   int n;
    int g, s, b;
    for(n=100;n<1000;n++)
    {   b=n/100;
        s=(n%100)/10;
        g=n%10;
        if(n==g*g*g+s*s*s+b*b*b) printf("%d\n", n);
    }
    return 0;
}
```

(2)

```
int wanbeishu()
{   int i,j,sum;
    for(i=1;i<=30000;i++)
    {   sum=0;
        for(j=1;j<i/j;j++)
            if(i%j==0) sum+=j+i/j;
        if(sum%i==0) printf("%d\n",i);
    }
    return 0;
```

}

(3)

```c
#include<stdio.h>
int main()
{   int ysSum(int x);                      /*函数原型声明*/
    int temp,i;
    for(i=1;i<=30000;i++)
    {   temp=ysSum(i);                     /*调用函数ysSum()计算i的约数和*/
        if(temp>30000) continue;           /*temp不能大于等于30000*/
        if(ysSum(temp)==i)                 /*如果temp的约数和等于i*/
            printf("\t%d    %d\n",i,temp);
    }
    return 0;
}
int ysSum(int x)                           /*函数定义*/
{   int i,sum=0;
    for(i=1;i<x;i++)
        if(x%i==0) sum+=i;
    return sum;
}
```

(4)

```c
#include<stdio.h>
float index(int n);                        /*函数原型声明*/
void main()
{   int number;
    printf("Please input element numbers(number ):");
        scanf("%d",&number);
    printf("The sum is %9.6f\n",index(number));
}
float index(int num)                       /*函数定义*/
{   int cnt;
    float temp,a=2,b=1,s=0;
    for(cnt=1;cnt<=num;cnt++)
    {   s=s+a/b;
        temp=a; a=a+b; b=temp;
    }
    return(s);
}
```

(5)

```c
#include<stdio.h>
long sum(int a,int b);                     /*函数原型声明*/
void main()
```

```
{   printf("1---40相加=%ld,1---80相加=%ld,1---100相加=%ld",
            sum(1,40),sum(1,80),sum(1,100));
}
long sum(int a,int b)                    /*函数定义*/
{   long x=0;
    for(int i=a;i<=b;i++)
        x+=i;
    return x;}
```

(6)
```
#include<stdio.h>
void fun(int n)                          /*函数定义*/
{   if(n/10) fun(n/10);
    printf("%d ",n%10);
}
int main(){
    int a;
    scanf("%d",&a)
    if(a>0) fun(a);
    printf("\n")
    return 0;
}
```

(7)
```
#include<stdio.h>
int sushu(int m);                        /*函数原型声明*/
void main()
{   int i,m;
    for(i=6;i<=1000;i+=2)
    for(m=3;m<i;m++)
        if(sushu(m)&&sushu(i-m))
            {   printf("%d=%d+%d\t",i,m,i-m); break; }
}
int sushu(int m)
{   int i;
    if(m<2)return 0;
    for(i=2;i<m;i++)
        if(m%i==0)return 0;
    return 1;
}
```

(8)
```
#include<stdio.h>
void main()
{   int hcf(int,int);                    /*函数声明*/
```

```
    int lcd(int,int,int);              /*函数声明*/
    int u,v,h,l;
    scanf("%d,%d",&u,&v);
    h=hcf(u,v);                        /*调用函数*/
    printf("h.c.f=%d\n",h);
    l=lcd(u,v,h);                      /*调用函数*/
    printf("l.c.k=%d\n",l);
}
int hcf(int u,int v)                   /*函数定义*/
{   int t,r;
    if(v>u){ t=u;u=v;v=t; }
    while((r=u%v)!=0)
    {   u=v; v=r;   }
    return(v);
}
int lcd(int u,int v,int h)             /*函数定义*/
{ return(u*v/h); }
```

第 7 章 指 针

1. 选择题

BDDCC　DCBBA

2. 填空题

(1) ① t=a[0]　　② j=1;j<n;j++

(2) ① y=d[0];　　② f(b,5,-1)

(3) ① *p-'0'　　② *(++p)

(4) ① *r<*p　　② r!=p　　③ p+1

(5) ① float x,float *a,int n　　② t=t*x　　③ return y

(6) ① i<=num/2　　② *(a+i)

3. 阅读下列程序,写出程序执行结果

(1)

0 1 2 3 4 5 6 7 8 9
1 2 3 4 5 6 7 8 9 10

(2)

1
5
7

(3)

0　　　0
1　　　2
2　　　4

0　　　0
2　　　3
4　　　6

(4)

1526

(5)

xyzABCabcd

(6)

21

(7)

191991919

本题中如果把 while 语句的循环体修改为如下：{t=*stb;*stb++=*ste;*ste--=t;}程序执行的结果：987654321

(8)

12345

(9)

13 -3

(10)

  1 2 3
 10 12 13

4. 编程题

(1)

```
#include<stdio.h>
char * first_unit[]={"zero","One","Two","Three","Four",
                    "Five","Six","Seven","Eight","Nine","Ten"};
char * second_unit[]={"Eleven","Twelve","Thirteen","Fourteen",
        "Fifteen","Sixteen","Seventeen","Eighteen","Nineteen","Twenty"};
void main()
{   char * change(int n);                    /*函数声明*/
    int num;
    printf("输入要转换的数:");
    scanf("%d",&num);
    if(num<0||num>20)printf("数字超出范围了\n");
    else printf("%s\n",change(num));
}
char * change(int n)                         /*函数定义*/
{   if(n>=0&&n<=10)return first_unit[n];
    else return second_unit[n-10-1];
}
```

(2)

```
#include<stdio.h>
void main()
{   char a[]="Language", *p=a;
    for(; *p!='\0'; p+=2)
        printf("%s\n", p);                   /*或 puts(p);*/
}
```

(3)
```c
#include<stdio.h>
void input(int num[ ])                          /*函数定义*/
{   int i;
    printf("请输入10个整数:");
    for(i=0;i<=9;i++)
        scanf("%d",&num[i]);
}
void mmx(int t[ ])                              /*函数定义*/
{   int  *n, *m, *p, *q,s;
    q=t+10;
    n=m=t;
    for(p=t+1; p<q; p++)
        if(*p> *n) n=p;
        else if(*p< *m) m=p;
    s=t[0];    t[0]=*m;    *m=s;
    s=t[9];    t[9]=*n;    *n=s;
}
void  output (int t[ ])                         /*函数定义*/
{   int  *p;
    for(p=t;p<=t+9;p++)
        printf("%d ", *p);
    printf("\n");
}
void main()
{   int num[10];
    input(num);
    mmx(num);
    output(num);
}
```

(4)
```c
#include<stdio.h>
int a[10];
void main()
{   int m=0,n;
    int fun(int x,int *y);
    for(n=100;n<=999;n++)
        if(fun(n,a)&&fun(2*n,a+3)&&fun(3*n,a+6))
            printf("No.%d: %d %d %d\n",++m,n,2*n,3*n);
}
int fun(int x,int *y)
{   int *p1,*p2;
    for(p1=y;p1<y+3;p1++)                       /*判断3位数字是否相同*/
```

```
        {   *p1=x%10;
            x/=10;
            for(p2=a;p2<p1;p2++)
                if(*p1==0||*p2==*p1)return 0;
        }
        return 1;
}
```

(5)

```
#include<stdio.h>
#include<stdlib.h>
void main()
{   int a, d, i, x,s=0,*p;
    p=(int *)malloc(200*sizeof(int));        /*开辟动态存储空间用于保存对称数*/
    for(i=10;i<=1993;i++)                    /*找对称数*/
    {   a=i; x=0;
        while(a!=0)
        {   d=a%10;
            x=x*10+d;
            a=a/10;
        }
        if(x==i) {   p[s]=x;   s++;   }
    }
    for(i=0;i<s;i++)                         /*输出对称数*/
       printf("%d\t",p[i]);
    printf("\n");
}
```

(6) 程序中根据 n 值的不同调用不同的函数。指向函数的指针变量还可以作为函数的参数,以便把函数的入口地址传递给另一个函数。当函数指针所指向的目标不同时,就可以调用不同的函数,且不需要对函数体做任何修改。

```
#include<stdio.h>
float   f1(float x)                          /*函数 f1*/
{   return (x);            }
float   f2(float x)                          /*函数 f2*/
{   return (2*x*x-1);      }
float   f3(float x)                          /*函数 f3*/
{   return (4*x*x*x-3*x);        }
float   f4(float x)                          /*函数 f4*/
{   return (8*x*x*x*x-8*x*x+1);}
void main()
{   float (*fp)(float),x,result;
    int n;
    printf("Input x and n:");
    scanf("%f,%d",&x,&n);
```

```
    switch(n)
    {  case 1:   fp=f1;   break;              /*为指针变量赋值*/
       case 2:   fp=f2;   break;
       case 3:   fp=f3;   break;
       case 4:   fp=f4;   break;
       default:  printf("Data error!");
    }
    result=fp(x);                             /*调用函数*/
    printf("Result=%f\n",result);
}
```

(7)

```
#include<stdio.h>
void main()
{   int i=0;
    char s[40], *sp;
    gets(s);
    sp=s;
    while(*sp)
    {   if(i%2==0) *sp='*';
        putchar(*sp);
        sp++;
        i++;
    }
    putchar('\n');
}
```

第8章 结构及其他

1. 选择题

DBCDC DBAB

2. 程序阅读,写出运行结果

(1) 12

(2) 51,60,21

(3) 10,x

(4) 264.00

(5) 0
    15.500000

(6)

Word value:1234
High value:12
High value:34

3. 编程题

(1)~(3)参考答案:

```
#include<stdio.h>
struct worker{
    char no[10];
    char name[20];
    char sex;
    int age;
    int workyear;
    float salary;
    char address[20];};
void main(){
    struct worker w1;
    printf("请输入该职工的相关资料:");
    scanf("%s,%s,%c,%d,%d,%f,%s",w1.no,w1.name,&w1.sex,
            &w1.age,&w1.workyear,&w1.salary,w1.address);
    printf("该职工的相关信息为:职工号:%s,姓名:%s,性别:%c,年龄:%d,
            工龄:%d,工资:%f,住址:%s",w1.no,w1.name,w1.sex,w1.age,
            w1.workyear,w1.salary,w1.address);}
```

(4)~(5)参考答案:

```
#include<stdio.h>
#define N 4
```

```c
struct stu                              /*定义关于学生信息的结构*/
{   int num;
    char name[20];
    int age;
    int score[3];
    float aver;};
void count(struct stu *s)               /*三门课成绩求平均值*/
{   int i,sum=0;
    for(i=0;i<3;i++)
    {   sum=sum+s->score[i];
        s->aver=sum/3.0;                /*将平均值存入相应地结构成员中*/
    }
}
void main()
{   int i,k;   float max;
    struct stu st[N],*p;
    p=st;
    printf("请输入%d个学生的信息:\n",N);
    for(i=0;i<N; i++)
    scanf("%d%s%d%d%d%d",&st[i].num,st[i].name,&st[i].age,
            &st[i].score[0],&st[i].score[1],&st[i].score[2]);
    for(i=0;i<N; i++)
    {   p=&st[i];
        count(p);
    }
    printf("输出学生的信息和平均成绩: \n");
    for(i=0;i<N; i++)
        printf("%d %s %f\n",st[i].num,st[i].name,st[i].aver,
    max=st[0].aver;
    for(p=st, i=0; p<st+N; p++, i++)
        if((p->aver)>max){k=i; max=p->aver;}
    printf("平均分最高者姓名、成绩是: %s   %f\n",st[k].name, st[k].aver);
}
```

(6)

```c
#include<stdio.h>
void main()
{   enum color{red,yellow,blue,white,black};
    int i, j, k, pri;
    int n, loop;
    n=0;
    for(i=red;i<=black;i++)
       for(j=red;j<=black;j++)
          if(i!=j)
          {  for(k=red;k<=black;k++)
```

```c
            if((k!=i)&&(k!=j))
            { n=n+1;   printf("%4d",n);
              for(loop=1;loop<=3;loop++)
              { switch(loop)
                  { case  1:   pri=i;    break;
                    case  2:   pri=j;    break;
                    case  3:   pri=k;    break;
                    default: break;
                   }
                switch(pri)
                { case red: printf("%10s","red");break;
                  case yellow:printf("%10s","yellow");break;
                  case blue:printf("%10s","blue");break;
                  case white:printf("%10s","white");break;
                  case black: printf("%10s","black");break;
                  default:break;
                }
              }
           printf("\n" );
          }
       }
    printf("\ntotal:%5d\n", n );
 }
```

第 9 章 文 件

1. 单选题
AADBA　CDBAA

2. 填空题

(1) ① 文本　　　② 文件结束

(2) ① 结束　　　② 1　　　③ 0

(3) ① fseek(文件指针,0,2)　② rewind 或 fseek(文件指针,0,0)
　　③ fseek(文件指针,10,0)

(4) ① fopen　　② w　　③ fclose(fp);

(5) ① "r"　　② "w"　　③ fgetc(p1)

(6) ① ==NULL　② !=NULL　③ strlen(s)-1

3. 阅读下列程序,写出程序执行结果

(1) ABCDEFGHI

(2) 从键盘连续输入 10 个字符,在输入字符的同时如果检测到输入的字符为大写英文字母则改为小写字母并写入文件 f1.dat,否则,直接写入该文件。

(3) First Second Third Four

(4) 把字符个数小于 20 的字符串写入到文件 t.txt 中,再从文件中读出后显示在屏幕上。

(5) 下列程序执行时输入 6↙,则输出结果是什么?

6,7,8,

4. 编程题

(1) 本题要求 3 位数的平方和等于 2000,所以这 3 个数字都小于 50。

```
#include<stdio.h>
#include<stdlib.h>
void main()
{   FILE * fp;
    int i,j,k,s=0;
    if((fp=fopen("f3.dat","w"))==NULL)
    {   printf("File open error.\n"); exit(0); }
    for(i=0;i<50;i++)
      for(j=0;j<50;j++)
        for(k=0;k<50;k++)
          if(i*i+j*j+k*k==2000)
          {   printf("%d,%d,%d\n",i,j,k);
              s++;
              fprintf(fp,"%d %d %d ",i,j,k);
```

```
        }
        printf("s=%d\n",s);
        fprintf(fp, "%d",s);
        fclose(fp);
}
```

(2)

```c
#include<stdio.h>
#include<stdlib.h>
void main()
{   FILE *fp;
    int i,d[10];
    if((fp=fopen("f2.dat","w+"))==NULL)
    {   printf("File open error.\n");  exit(0);  }
    for(i=0;i<10;i++)
    {   scanf("%d",&d[i]);
        fprintf(fp, "%d ",d[i]);
    }
    fclose(fp);
}
```

(3) 从文件中读取数据时,应与写入时的格式相同,否则,读取的数据会出错。

```c
#include<stdio.h>
#include<stdlib.h>
void main()
{   FILE  *fp;
    int d[10],i;
    if((fp=fopen("f2.dat","r"))==NULL)
    {   printf("File open error.\n");  exit(0);  }
    for(i=0;i<10;i++)
    {   fscanf(fp, "%d ", &d[i]);
        printf("%d ",d[i]);
    }
    fclose(fp);
}
```

(4)

```c
#include<stdio.h>
struct student
{   int num; char name[20]; int age; char sex; int score[5];};
void main()
{   struct student st[10];
    FILE *fp;
    int i, j;
    for(i=0;i<10;i++)
    {   scanf("%d %s %d %c", &st[i].num, st[i].name, &st[i].age, &st[i].sex);
```

```c
        for(j=0;j<5;j++)
            scanf("%d",&st[i].score[j]);
    }
    if((fp=fopen("s.dat","w"))==NULL)
    {   printf("Cannot open.\n"); exit(0);   }
    for(i=0;i<10;i++)
        fwrite(&st[i], sizeof(struct student), 1, fp);
    fclose(fp);
}
```

(5)

```c
#include<stdio.h>
struct st
{   int num;char name[20];int age;char sex;int score[5];}s[10];
void main()
{   FILE *fp;
    int i, j;
    fp=fopen("s.dat","r");
    for(i=0;i<10;i++)
        fread(&s[i],sizeof(struct st),1,fp);
    fclose(fp);
    for(i=0;i<10;i++)
    {   printf("no:%d, name:%s, age:%d, sex:%c,",s[i].num, s[i].name, s[i].age,
        s[i].sex);
        printf("score:");
        for(j=0;j<5;j++)
            printf("%d ",s[i].score[j]);
        printf("\n");
    }
}
```

(6)

```c
#include<stdio.h>
#include<stdlib.h>
void main()
{   int i;
    FILE *fp=fopen("f4.dat","w");
    if(fp==NULL){   printf("File open error.\n"); exit(0); }
    for(i=0;i<32767;i++)
    if(i%3==1&&i%5==3&&i%7==5&&i%9==7)
    {   printf("%d\n", i);
        fprintf(fp, "%d ", i);
        break;
    }
    fclose(fp);
}
```

(7)
```c
#include<stdio.h>
#include<stdlib.h>
void main()
{   int i, j, k, a[100], sum=0;              /*a数组用来存放每个数的因子*/
    FILE *fp=fopen("f5.dat","w");
    if(fp==NULL) { printf("File open error.");exit(0); }
    for(i=6;i<=1000;i++)
    { sum=0;                                  /*sum用来存放因子之和*/
        for(k=1,j=0;k<i; k++)
            if(i%k==0) { a[j]=k; j++; }       /*找出各因子存放到数组中*/
        for(k=0;k<j; k++)
            sum=sum+a[k];                     /*计算各因子之和*/
        if(i==sum) {  printf("%d\n",i); fprintf(fp,"%6d",i); }
    }
    fclose(fp);
}
```

(8)
```c
#include<stdio.h>
#include<stdlib.h>
void main()
{   int p=0,n=0,z=0,t,sum,num,x;
    FILE *fp;
    printf("Please input total:");
    scanf("%d",&sum);sum
    fp=fopen("n.dat","w+");
    if(fp==NULL) printf("fole open error");
    else
    {   printf("Input integers:");
        for(t=0;t<sum;t++)
        {   scanf("%d",&num);
            fprintf(fp,"%d ",num);
        }
    }
    rewind(fp);
    while(!feof(fp))
    {   fscanf(fp,"%d ",&x);
        if(x>0) p++;
        else if(x<0) n++;
        else z++;
    }
    fclose(fp);
    printf("positive:%3d,negtive:%3d,zero:%3d\n",p,n,z);
}
```

第 10 章 编译预处理与位运算

1. 选择题
CCBAB BCABA AD
2. 编程题
(1)

```c
#include<stdio.h>
#define  swap(x,y) {int t;t=x;x=y;y=t;}
void main()
{   int i,a[10],b[10];
    printf("输入 a、b 数组的值。a 数组是：\n");
    printf("a 数组是：");
    for(i=0;i<10;i++)
        scanf("%d",&a[i]);
    printf("b 数组是：");
    for(i=0;i<10;i++)
        scanf("%d",&b[i]);
    for(i=0;i<10;i++)
        swap(a[i],b[i]);
    printf("交换后 a、b 数组的值分别是：\n");
    printf("a 数组是：");
    for(i=0;i<10;i++)
        printf("%d ",a[i]);
    printf("\nb 数组是：");
    for(i=0;i<10;i++)
        printf("%d ",b[i]);
}
```

(2) 求两个整数相除的余数。

```c
#include<stdio.h>
#define SURPLUS(a,b)   ((a)%(b))
void main()
{   int a,b;
    printf("请输入两个整数 a,b:");
    scanf("%d%d",&a,&b);
    printf("a,b 相除的余数为:%d\n",SURPLUS(a,b));
}
```

(3)

```c
#include<stdio.h>
```

```c
#define LEAP(year) (year%4==0)&&(year%100!=0)||(year%400==0)
void  main()
{   int year;
    printf ("Please input a year :");
    scanf ("%d",&year);
    if(LEAP(year))
        printf ("%d is Leap Year.\n",year);
    else
        printf("%d is not Leap Year.\n",year);
}
```

(4)

```c
#include<stdio.h>
#define TOUPPER(ch) ch-32
#define TOLOWER(ch) ch+32
void  main()
{   int i=0;
    char str[20];
    gets(str);
    while(str[i]!='\0')
    {   if(str[i]>='a'&&str[i]<='z') str[i]=TOUPPER(str[i]);
        else  if(str[i]>='A'&&str[i]<='Z') str[i]=TOLOWER(str[i]);
        else ;
        i++;
    }
    printf ("%s\n",str);
}
```

(5)

```c
#include<stdio.h>
void  main()
{   int n,s,k;
    printf ("请输入一个整数 s 和循环移位的位数 n:\n");
    scanf("%d%d",&s,&n);
    k=8*sizeof(unsigned);
    s=s<<n|s>>(k-n);
    printf ("%d\n",s);
}
```

参 考 文 献

[1] 谭浩强.C 程序设计题解与上机指导(第三版).北京：清华大学出版社,2005.
[2] 谭浩强.C 程序设计试题汇编.北京：清华大学出版社,1999.
[3] 刘光蓉.C 程序设计实验与实践教程.北京：清华大学出版社,2011.
[4] 黄维通.C 语言程序设计习题解析与应用案例分析(第 2 版).北京：清华大学出版社,2010.
[5] 张建宏.C 语言程序设计实践教程.北京：清华大学出版社,2009.
[6] 姚合生编著.C 语言程序设计习题集、上机与考试指导.北京：清华大学出版社,2008.
[7] 林小茶编著.C 语言程序设计习题解答与上机指导.北京：中国铁道出版社,2007.
[8] 牛志成编著.C 语言程序设计习题与上机指南.北京：清华大学出版社,2008.
[9] 黄明,梁旭,万洪莉编著.C 语言课程设计.北京：电子工业出版社,2006.
[10] 盛夕清,赵阳,林科学编著.C 语言程序设计学习指导、实验指导与课程设计.北京：中国水利水电出版社,2006.
[11] 颜晖主编.C 语言程序设计实验指导.北京：高等教育出版社,2008.
[12] 张引.许端清等编著.C 程序设计基础课程设计.杭州：浙江大学出版社,2007.
[13] 姜雪,王毅,刘立君编著.C 语言程序设计实验指导.北京：清华大学出版社,2009.
[14] 周彩英主编.C 语言程序习题解答与学习指导.北京：清华大学出版社,2011.
[15] 王敬华,林萍,张维编著.C 语言程序设计教程习题解答与实验指导.北京：清华大学出版社,2006.
[16] 陈朔鹰,陈英,乔俊琪编著.C 语言程序设计习题集.北京：人民邮电出版社,2000.
[17] 常东超,吕宝志,郭来德等编著.C 语言程序设计.北京：清华大学出版社,2010.
[18] 黄维通,马力妮编著.C 语言程序设计.北京：清华大学出版社,2005.
[19] 崔武子,林志英,和青芳编著.C 程序设计案例教程及题解.北京：清华大学出版社,2010.
[20] 裘宗燕编著.从问题到程序 程序设计与 C 语言引论第 2 版.北京：机械工业出版社,2011.